JOHNSTOWN WATERS

JOHNSTOWN WATERS

THE LEGACY OF THE STEEL INDUSTRY

JOSHUA PENROD

THE History PRESS

Published by The History Press,
An imprint of Arcadia Publishing
Charleston, SC
www.historypress.com

First published 2025

Manufactured in the United States

ISBN 9781467159524

Library of Congress Control Number: 2025931876

For the people of Johnstown and Somerset,
my parents, my family, my friends.

CONTENTS

ACKNOWLEDGEMENTS

With great gratitude and acknowledgement of the staff and leadership, present and former, of the Cambria-Somerset Authority, including Rick Ames, Earl Waddell, James Greco and Kathleen. A deep thanks, as before, to Richard Burkert, former president of the Johnstown Area Heritage Association.

With great appreciation as well to Professor Albert Churrella, author of the definitive multivolume history of the Pennsylvania Railroad, who came in with some timely and critical information.

And a hearty thank-you to my colleague Rachel Liebe, who helped enormously with photo preparation.

Thank you to my parents for providing comments on earlier drafts of this work.

Any and all mistakes are my own.

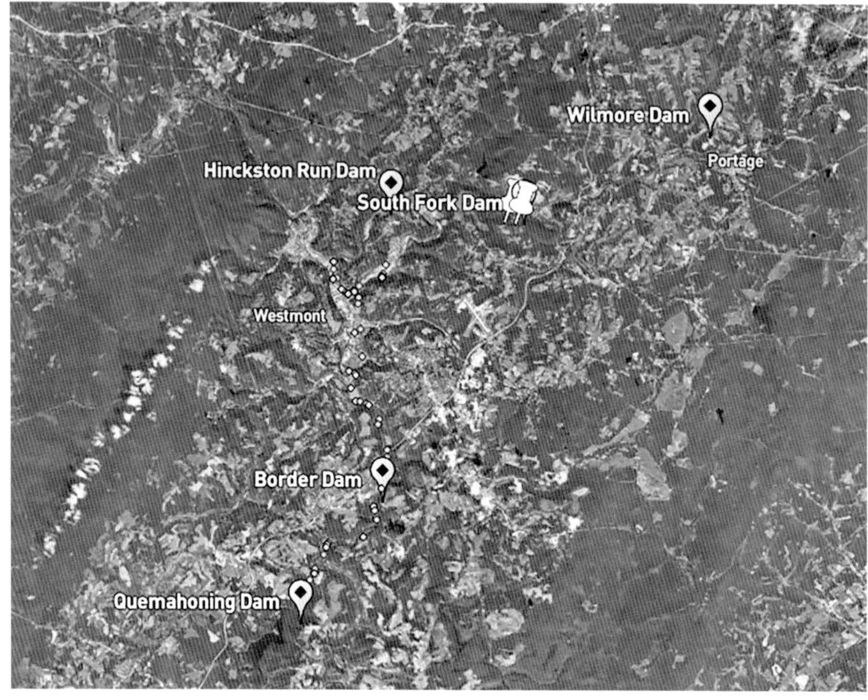

A Google Earth image showing the map of the dams and reservoirs in the Cambria-Somerset Authority's technonatural system. *Courtesy CSA.*

INTRODUCTION

WHY HERE? AND—DEFINING THE SYSTEM

Steel: A Poem
As the molten steel leaves the huge converter
Like the steady stream from a surging heart
The future of its products looms ahead.
Will these ingots become strong rails—on which the fastest limited shall ride;
Pullman cars, in which eager travelers will sojourn, enjoying the panorama before them;
Or an engine, as important to transportation as beauty to a poet?
Every particle may enter into the making of a massive steamboat, seen the world over;
Or into the numerous cogs of automobile wheels, traversing the highroads and by-roads;
Or, perhaps, it will become a part of many window sashes
proudly adding their help to construction,
In whatever commodity, the momentous renditions of service
Make the molten steel God's gift to mankind.

—Ruth Wisor, graduate, class of 1935, Johnstown High School

To say that Johnstown, Pennsylvania, has an uneasy relationship with water is to spell perhaps *the* understatement of industrial history. Virtually everyone with a passing acquaintance knows at least the gist of the story: in 1889, more than 2,200 people lost their lives in a sudden flood, when the South Fork dam, a dozen miles upstream from Johnstown, breached and billions of gallons of water tore southbound through the valley leading directly to the city, destroying buildings, people, animals and everything else in the torrent's path. The story has been made into sensationalistic exploitation depictions, along with serious documentaries and scholarly work. The rich owners of the South Fork Fishing and Hunting Club, including notables such as Andrew Carnegie, escaped from any true consequences of what many would readily call negligence.

Even so, digging coal, making steel and driving locomotives all have one thing in common: the need for water. The water must be ready in vast quantities, delivered on demand to a complex and growing enterprise. Any such firm must then figure out how to do this, in a town where a massive disaster was fresh in the living memory of the thousands left behind after the damage, these same thousands also in mourning and burying their dead. How to balance the obvious need for reliability and safety, the public perception of the same, and then *actually* create the system that will supply the water to a vast, growing and complex manufacturing system? This is the difficult balancing act that Cambria Steel (re-branded and re-incorporated from Cambria Iron in the late 1800s) sought to undertake.

The geography around Johnstown, while undoubtedly being of great importance in considering the great damage of the 1889 flood, also creates and limits the engineering possibilities of designing and building a water supply. At the same time, it is the natural history of the area that created and limited the possibilities of heavy industry; the area is blessed with transportation options and natural resources such as wood, coal and, in the early days, ore. By the early 1900s, Cambria Steel was an integrated steel works with a stunningly wide range of production of types of steel. A leather-bound product catalogue from the company published in 1909 gives a snapshot of the impressive capacities of that company. With offices in New York, Chicago, Tacoma, San Francisco and several other cities spread around the United States, its corporate headquarters in Philadelphia and its "works" in Johnstown, the company offered a bevy of raw and finished steel products. The line included structural steel, train cars, pins, forgings, axles, wire, rails, bar steel and even seats for tractors and other agricultural applications. The catalogue also notes the different processes that Cambria would use, specific to each application and the requirements of its customers, and enumerates how the company would also adhere to the updated standards of the industry of the time.

At the same time, Cambria was not only an integrated steel works but also had its own coal mines in the area; such an integration is likely what made the eventual purchase of the company by Midvale Steel and, ultimately, by Bethlehem Steel, an attractive one. The challenges were a necessary part

An upstream view of Hinckston's Run, leading into the reservoir of the same name. *Author's collection.*

of the advantages, however, with steep valley walls, heavy forest and any number of creeks and streams winding and twisting into the major rivers in the area. The last component in the challenge column obviously has to be the weather itself; western Pennsylvania is not generally known as a mild climate, with significant winters creating long periods of ice, snow and subfreezing temperatures.

As we see in this work, making steel requires water, and lots of it. The system of "industrial water supply" first came about by Cambria Steel beginning in the second half of the 1800s. It evolved and grew over time, and the system described in this book is quite specific, looking at several dams and reservoirs out of many in the region.

The system as described here and as it still exists today was largely built and completed by 1913. In essence, the system consists of five dams with three reservoirs, with the smaller two dams holding retention pools in rivers rather than being proper reservoirs. The three main dams and reservoirs—the Quemahoning (12 billion gallons), Wilmore (1 billion gallons) and Hinckston Run (1 billion gallons)—were all owned by Bethlehem Steel at

Construction and reinforcement of Wilmore Dam, circa 1960. *Courtesy CSA.*

the time of that company's final days.[1] The entirety of the system spans about 550 square miles of drainage area, with 1,200 total acres of water and 4,000 acres of land surrounding the reservoirs. It is once again important to distinguish the impoundment dams along the riverways—the Border and South Fork Dams—as they are part of the regulation of the system, but the water from the Quemahoning (oftentimes shortened to the Que), Wilmore and Hinckston Dams is the water that is the primary resource as used. The total pipeline used by the system is over 344,000 feet, or nearly 60 miles. The Wilmore Dam was purchased by Bethlehem Steel in 1960, but the other dams were constructed and maintained by Bethlehem or its predecessor, the Cambria Steel Company. A subsidiary of Cambria and Bethlehem, the Manufacturers Water Company (MWC), administered the system for around a century, and then it came under the joint ownership of two neighboring counties, Cambria County and Somerset County. The Cambria Somerset Authority, or CSA, is the current body charged with the maintenance, upkeep and operation of the system today, as of this writing. The original purpose of the system was to provide water for steelmaking purposes, but now the area is one of ecological significance, wildlife conservation and outdoor recreation.

What once was an industrial water well can now look a lot like a natural oasis. The absolutes are not, however, the reality—as they so rarely are. There has been plenty of outdoor recreation in the area since before its inception, and at the same time, the CSA continues to balance the need for conservation, economic opportunities and public and private interests.

While the 1889 flood has been covered in great detail in both film and the printed word,[2] it should do to recap it here at the outset, given the repeated and common reference to water as a resource, tool and destroyer and given the importance of the geography to the area and its heavy industrial history and heritage, along with the role that the geography itself played in the catastrophe that befell the city on Memorial Day weekend 1889. Beyond the aspects of disaster tourism, there are intricate and difficult consequences to the sociology and economy of the people in the area, many of whom are the descendants of the survivors of the 1889 flood and almost all of whom were affected, in some shape or form, by steel, coal and rail.

A number of rich industrialists of the time, including the aforementioned and well-known Andrew Carnegie, had ownership shares in a private hunting ground and water reservoir—Lake Conemaugh—at the South Fork of the Little Conemaugh River, about a dozen miles north and upstream of Johnstown. The South Fork Fishing and Hunting Club, as it was known, was a warm-weather retreat for the tycoons of Pittsburgh, given the higher elevations in the mountains and its location upwind from the prevailing breezes of the massive Johnstown-based steel facilities downriver. The dam at the location had been built decades earlier by the state to supply water to the Pennsylvania Canal system, which subsequently went out of service with the ascendancy of the Pennsylvania Railroad, which purchased the dam and reservoir and then sold the property to the private interests.

Due to heavy rains and general poor upkeep of the earth-filled dam, the structure broke, unleashing somewhere north of three billion gallons of water downstream, essentially all at once. The close confines of the valley walls channeling the Little Conemaugh caused the onrushing wall of water to gain in both height and speed as it tumbled downhill, scouring everything in front of it, wiping houses, trees, animals and people clean from the face of the earth. To a close observer, it would have seemed like nothing so much as a rapidly approaching roaring wall of mist, with a roiling mass of trees and structures tumbling over and over again before it.

A lone telegraph operator attempted to signal Johnstown, but the wires were broken. The signal made it partway, and brave runners on horseback raced ahead in an effort to save lives. Just north of Johnstown, an alert railroad engineer named John Hess sprinted into action, reversing the direction of his train, running it at full power *in reverse*, the whistles tied down to their maximum setting, louder than a modern jet engine, alerting everyone in any proximity to flee, for something was happening—something had gone terribly, terribly wrong. Only God knows how many lives Hess saved as people fled uphill, while Hess stayed in his locomotive as long as he possibly could, watching the onrushing wall of mist edging closer to him, outpacing the train, inexorably closing the gap to the nose of his backward-racing locomotive.

Hess escaped.

Much of the rest of the city did not.

Johnstown then, as it is today, is seated at the bowl of two valleys, the confluence of two rivers: the Little Conemaugh and the Stonycreek (sometimes also spelled Stoneycreek or Stoney Creek). The southbound flood entered the bowl with enough force to strip nearly every structure from its foundation. Tens of thousands of people were displaced in mere seconds, and well over two thousand perished in the flood and subsequent fire that ignited due to the dozens of fuel tanker cars that had overturned

The widespread destruction of the Johnstown flood of 1889, showing few buildings and churches remaining. *Courtesy Library of Congress (LOC).*

Another perspective on the aftermath of the 1889 flood. *Courtesy LOC.*

and been forced into that same bowl. There was plenty of tinder and fuel for the flames as houses, buildings, all sorts of debris such as barbed wire, structures, cattle and, of course, the terrified adults and children, the very young and the very old, who, saved from the initial destruction of the flood, were fated to be burned to death in an inferno that surged through the fuel floating about and atop the water. The water itself was pent up behind the great Stone Bridge of the Pennsylvania Railroad, just next to the great Cambria Iron plant. Everything carried by and floating in that water stopped behind that new, unanticipated dam, a state in which people had a choice of drowning, being crushed or burned to death. As it so happened, it wasn't so much the great Johnstown flood of 1889 as much as it was the great Johnstown flood *and* fire.

Eventually, the reckoning came to better than 2,200 dead, more than 700 of whom[3] lay unidentified in unmarked graves in Grandview Cemetery high on a hill above the city, where a magnificent sculpture watches over the silent and unidentified dead, the victims of negligence wrought by greed and sources of power above the reach of both law and justice.

As one might expect, there were significant recriminations in the aftermath of the disaster. Andrew Carnegie donated $10,000 (approximately $330,000 today) to rebuild the library in Johnstown; this building now houses the museum dedicated to the events of the 1889 flood and remembrance of the victims and families affected. Carnegie was lauded for the gesture, but others believed him and his wealthy ilk to be solely responsible for the disaster that claimed those thousands of lives. Legal doctrines involving strict liability for such a failure were instituted, as the South Fork Hunting and Fishing Club and its members were exonerated of all wrongdoing. The change in the laws, of course, benefited none of the actual victims.

It is the memory of this extraordinary calamity in the minds of so many in the area that, just a few years after the flood, prompted the protest of the Cambria Steel Company's construction of another large dam south of the city, along the Stonycreek River. Before we get there, however, we should define some terms, set the stage and better describe the system that is the subject of this book.

1

SETTING THE STAGE

A picture of the United States around 1900 lends greater context to not only Johnstown and its industrial footprint but also some of the "whys" behind actions taken. As an example, the short-lived President McKinley was in office, and the nation was 35 years removed from the close of the Civil War. The United States was engaging in trade negotiations with China and also involved in military patrol actions in Central America while fighting a colonial war in the Philippines at the same time. The Wright brothers initiated their flying experiments, and the Galveston Hurricane hit the Gulf Coast of Texas, killing somewhere in the neighborhood of 8,000 people, thereby becoming the deadliest disaster in the nation's history, superseding the Johnstown flood of 1889. Several major mining disasters also occurred, in West Virginia and in Utah, and the United States and Britain entered an agreement with Nicaragua for a canal to bridge the Caribbean Sea and Pacific Ocean. It does not strain the imagination, in a sense, to pause and consider whether things have changed overly much over the past century or more.

Leading up to 1900, the country had gone through an unprecedented social, economic, political and military expansion. With the quelling of the Civil War and the putative end of legal slavery, industrial concerns started dominating the economic sphere, rapidly overtaking agriculture as the nation's primary economic driver. The manufacturing started, of course, far earlier with textiles in New England and even prior to that in the United Kingdom, but by the late 19th and early 20th centuries, the dominant visible structures, particularly in the Northeast, were steel mills and railroads. The other dominant but unseen structures were the mines and likely that much larger for it.

So much of history is built on contingency. Entire academic careers have been made of examining a handful or perhaps even one. However, when we look back into deep time, the picture is one of great change that has taken place over eons, the true dimensions of which are essentially unfathomable. Pictures constructed from evidence dating from several hundred *million* years ago might be incomplete, and looking back that far might not immediately seem relevant; that, however, is too glib. The story of western Pennsylvania, Johnstown, the hills and valleys, soils and waters must begin with an underlying appreciation for its geology and geography. Dinosaurs and giant ferns might be the signature objects of the sparkling-eyed attention of children, the occasional paleontologist and the moviegoer who enjoys big-budget Hollywood cinema, but it's the very nature of nature itself that shaped the lives of the people and created the places explored in this work.

Some 300 million years ago, all the continents—the land masses from which all humanity hails—were jammed together into one supercontinent scientists call Pangea. This means exactly what it is—all of earth, entities we think of today as Antarctica, Asia, the Americas and everything else were together as one, with no Atlantic Ocean separating East from West. In this single large continent, the place we now call Pennsylvania was fairly close to the geographic center of Pangea and was located along the equator; the Appalachian Mountains themselves were formed at the point of collision between the northern and southern masses joining themselves into the supercontinent. At the same time, oxygen levels were far higher in the atmosphere then, and the world looked much different. These are the times before the dinosaurs made so famous by cartoon and movie alike, before there truly were any great land animals at all.

This section of deep geological time is known today as the Carboniferous period, and a subsection of this period is called, appropriately enough, the Pennsylvanian. Though the root of the word itself is obviously familiar, the planet then was as alien as any that might be seen on the silver screen today. It featured nightmarish millipedes two yards long and dragonflies with wingspans of three feet. The air had much higher levels of oxygen than today, around 70 percent more. Strange trees, more related to mosses and ferns than trees familiar to us today, covered the hot, wet landscape. They made for vast rain forests circling shallow seas, a world of wetlands.

Picture of a Pennsylvania deep coal mine, circa 1895. *Courtesy LOC.*

With climate cycles, billions of tons of organic matter would be flooded and compressed time and time again, resulting in the hard, burnable, carbon rock that today we call "coal." The carbon locked in hundreds of millions of years ago by this cycle created the fuel responsible, in large part, for the prosperity and unprecedented economic growth driven by heavy industry until now, as we may face new challenges created by this unlocked energy. It remains our legacy, still used to a substantial extent, that now presents us with the new challenge of pollution and mine drainage and enormous mounds of mining detritus, often called bony piles.

Over that deep time, Pangea once again broke apart, and millions of different species of plants and animals arose, peaked, declined and went extinct. Pennsylvania moved northward as its continental home shifted and turned, eventually creating the continent of North America. In large part, it stayed far enough south to avoid glaciation through the last glacial maximum only a few tens of thousands of years ago; waters cut through the eastern part of this continent, and time wore the Appalachians down from their maximum height (where they might well have looked a lot like the Rockies) and created a convoluted and complex agglomeration of hills and valleys, a wrinkled landscape fraught with crossing waters, steep hills and winding, torturous paths.

The water then became a highway, a natural part of the cultural infrastructure for the aboriginal population and then for European colonists, traders, conquerors and settlers. It never lost its importance as a muscle and a nerve, for slaking thirst, for driving agriculture, for communications and trade and for building heavy industry, a conception so deeply invested in different understandings and possible definitions of "infrastructure." Indeed, perhaps the most fundamental of all components of infrastructure in heavy industry is water. This is true, irrespective of the need for fuel, ore, knowledge—and also decidedly separate and apart from any social fear of it.

An extensive hydrological peer-reviewed study by Coughenour and colleagues noted some of the hydrology of the region that continues to make it at flood risk, even in the context of catastrophe. The 1889 flood, for example, while proximately caused by extraordinary levels of negligence, needs greater context of the history of flooding in Johnstown. Aside from the major floods of 1889, 1936 and 1977, Johnstown has had at least 47 floods since the early 1800s. The steepness of the valley in which the city is situated, along with the flow, volume and concentration of the Little Conemaugh, the Stonycreek and many other tributaries entering the system and concentrating water-force on the area all combine for a unique hydrological and geographic situation. "The rugged topography of the basin favors rapid runoff and delivery of floodwaters to confined valley floors." The authors make several important observations of note—that the hydrology of the area's rivers have not been adequately studied despite the extreme history; that had the South Fork Dam of 1889 not been modified by its keepers and owners, it would have been able to withstand the storm activity that precipitated the flood that killed thousands;[4] and flooding continued to plague Johnstown after the 1889 flood, with flood events occurring in the downtown area every year up until 1913. It was the vast channelization efforts post-1936 that brought this under control, and after 1972's Hurricane Agnes, they were thought to have converted Johnstown from flood-prone to flood-proof. This, of course, was shown to be nonsense with the cataclysm of the 1977 flood, but Coughenour also makes a strong argument that the 1972 Agnes event, which flooded vast parts of the Eastern United States and resulted in catastrophe along the Susquehanna basin, would have caused Johnstown another flood even with the channelization efforts, but Johnstown was just a bit farther west than the main track of that storm. The city ended up with somewhat less rainfall that would lead, five years later, to the 1977 flood, where an excess of eight inches of rain fell in 11 hours. They also note that the 1977 flood would have been far worse had the channelization efforts not been made. At least in the instance of 1972, Johnstown was on the right side of geography.

The mineral-rich area around Johnstown treated early industrial attempts with the availability of fuel—at first in terms of wood for charcoal but then later with enormous reserves of coal—along with ore and transport by rail and water. The Pennsylvania Canal, initiated in 1820, established a foothold in Johnstown and neighboring Portage with a system of trolleys and trains to pull canalboats on an inclined plane over the formidable Allegheny Ridge—much lower than its peak of millions of years before but a formidable barrier nonetheless. The railroad initiated in Hollidaysburg at the terminus of the eastern half of the canal carried cargo over the mountains first by donkey train and then by steam engine, descending to the eastern terminus of the western half of the canal in Johnstown.

This system, while becoming rapidly outdated and made obsolete by the burgeoning Pennsylvania Railroad, created just fertile enough possibilities to be a launching site for larger industrial firms and technical systems over the ensuing decades. Still, the geography would forever be a challenge for Johnstown. A bit outside of the main area of transport and communication between Pittsburgh and Philadelphia, New York, Washington and Baltimore, later owners of the steel concerns were faced with higher costs and difficult supply chain and logistics in Johnstown, as fine an example as both path dependence and various interplays between social pressures and economic realities that might be observed. The only time the Johnstown mills were state-of-the-art was in the later part of the 19th century; for the rest of the 100-plus-year history, the plants were, at best, on par with other facilities and, more usually, less efficient and higher cost.

Midvale Steel acquired Cambria Steel in 1915, and the massive Bethlehem Steel Company subsequently acquired Midvale in 1923. Even since 1889, industrial thirst for water rarely abated, and the system supplied an embarrassment of riches to slake that thirst. Despite the death and destruction of the 1889 flood, the Cambria Steel Company passed MWC to Midvale and Bethlehem, which remained in operation through the turn of the 20th century to its end around the year 2000. MWC built what would become for some time the largest dam and water supply in the state, at the Quemahoning Creek. By the time Bethlehem started its inevitable collapse in the 1970s and into the 1980s, three large reservoirs and two smaller ones supplied the industrial water to the steelmaking facilities, mines and contracted partners throughout the region.

What Is a Dam?

Dams take many forms. The giant dams of the New Deal era are perhaps the most famous national examples (except perhaps for the dam that failed Johnstown in 1889), but they've also long been a part of humanity's interaction with the natural world. After all, few things have been more crucial throughout human history than the ability to control and direct water, especially for flood safety, defense and agriculture. Channeling water to grow crops and tend to livestock were of great importance to any semi-centralized state, as agriculture might be recognized as a cornerstone—if not *the* cornerstone—of concentrated state power. To control water is to control life.

A dam is a structure erected across a body of moving water, be it a creek, stream or river. In more recent times, these structures can be made of steel-reinforced concrete and are highly engineered machines, but for thousands of years, humans made do with what was at hand—soil, clay, rocks and debris. There is evidence of dam construction as far back as 5,000 BP in the Middle East; other examples exist in China that date back even further. The Romans too were inveterate dam builders. Other enterprising sorts in Sri Lanka, Japan and India also built dams as a way of manipulating nature to benefit human needs, and it is likely that historians and archaeologists will continue to find many examples of ancient dam sites. They leave a lasting mark.

While this work offers greater detail as to the dams in the system, we can note at the outset that there are a number of time-tested ways to construct a dam, particularly today with the usage of highly specific types of concrete and other materials, but earth-filled dams are generally the most common around the world, predating the use of concrete and still in use today. Of course, the famous dam that failed Johnstown in 1889 was an earth-filled dam, consisting (at first) of clay soil, covered in rocks known as rip-rap and accompanied by spillways, which are directed channels over which water can flow should the level of water in the reservoir exceed the amount that might be safely held by the dam. Despite the 1889 failure of the South Fork Dam, this is not an indictment of that design; it is instead an indictment of the mismanagement and indeed, negligent malfeasance wrought by the owners of that dam, who altered the dam beyond its safety parameters, such as by raising the spillways and preventing them from serving their purpose, and by the installation of a fish weir, which, while meant to corral sport fish in the reservoir and stop them from escaping downstream, also had the effect of collecting greater debris, which exerted more and more pressure on the breast of the dam. It was only a matter of time until failure occurred. This was not unknown at the time; it was just a calculation that retaining trout to catch was more important than human lives. We should note this here, as it is a critical recognition of the importance of maintenance in infrastructure in a technical system—not just avoiding inconvenience but also guarding against miscalculation and sometimes out-and-out bungling.

The ancient rivers and hills of Pennsylvania combine to create possibilities to be exploited by human intelligence and engineering, however, and the usage of water for industrial purposes is just as critical as it is for agricultural. We also note throughout this work the existence of a "techno-natural" system, the frank and obvious merger of human manipulation and nonhuman, natural structure. The same geographical accident that created the combination in which investors and industrialists sought to site steel mills and coal mines also created the means by which that water could supply those same industries. In the case of the dam-and-reservoir system around the Johnstown area, the combination of gravity and eons created the background conditions under which the possibilities were created and those under which operations could be continuous.

An earthfill dam relies on a core of impervious materials, mostly clay soils.[5] The Johnstown area has various types of soils as well, different types of which are associated with different environmental exhibits (certain types of soils are associated with woodlands, for example, or are better for crops). The soils contributing to the basis of the Conemaugh and Stonycreek Rivers, and the basin in which the city itself sits, comprise a poorly drained set of clay soil types called the Monongahela, Whitwell and Tyler soils. These prove poor for agriculture, but true to the geographic convergence present in this area, the local clay soils are an important component in constructing a large earthfill dam.

With a clay core established, an earthfill dam can either consist of one type of material or be of a multitude of layers of different materials. The bases of such dams are usually very wide, far wider than the dam is tall. They are dams that are adaptable to challenging environments, particularly those built (as these in western Pennsylvania) prior to the development of the engineering of towering concrete and steel dams, such as Hoover. Nonetheless, the earthfill dams in the Johnstown-area system hold and have held significant amounts of water for more than a century now. They've played a nearly silent but also irreplaceable role in the history of the area and, in so doing, in the history of the country.

More than a century since their construction, these dams have not failed. This is not to say they are indestructible, of course—for, with the exception of gold or diamonds, everything deteriorates over time. This basic truth brings us to the other consideration involved with infrastructure, which is its continued need for observation, monitoring and maintenance despite their original purpose, owner and controller fading into history. As safe as these dams have been, there is no question that humans must still and continuously intervene in their creation to ensure that they remain viable.[6]

Another thread runs through the history of the dams: the thread of conservation and environmental concern. Certainly one of the plainest levels of awareness, common to virtually everyone, is the environmental impact that heavy industry can have on its surrounding areas. Some places are better than others, of course, but the entire region of western Pennsylvania is thoroughly pockmarked and scarred by deep and strip mines, brownfields and the deteriorating physical plants of the many steel and related mills and foundries and hillsides and hilltops removed for extraction or choked off by smoke and denuded of trees and other plants, much less any animal life that would have been supported by the vast chestnut forests of pre-European contact in North America. Streams, creeks and rivers still to this day often flow orange and red with iron, sulfur and other runoff generated by mining activities. While many have been and continue to be improved and sometimes restored, much room for improvement remains.

Another theme explored with the dams and reservoirs is through the analytical tool of the Large Technical System, or LTS, a concept largely developed by the noted University of Pennsylvania historian of technology Thomas Hughes. Since expanded and elaborated on by numerous scholars, the LTS provides an idiom in which many social factors are enrolled in combination with built technologies and physical artifacts. More specifically, the social factors include economic and political constraints, exchanges of

Clear view of the masonry composition of the Wilmore Dam, circa 1960. *Courtesy CSA.*

information and knowledge, interactions of labor and management and many other related inputs that combine to form the system. The result is an integrated combination of the physical, the cultural and many forms of knowledge.

In the same vein, another unifying narrative strand in the history of these dams and reservoirs is the conversion of the LTS from industrial into ecological. The documentation of the evolution is present and significant, though the integration of the LTS idiom and the narrative of change and evolution into a transformation still occurs today. However, given that at least some of the usage of the waters and surrounding land were ecological and leisure from the outset,[7] how much of the "transformation" is actual in the system itself, and how much is a general *perspective* that shifted through society's change over time—particularly from the 1980s, which was the decade of the precipitous and significant decline of the steel industry in Johnstown?

The one direct answer to such a challenge is the obvious shift in laws and policies that came into effect between the years of the construction of

Looking northward across the top of Hinckston Run Dam to the Reservoir. *Author's collection.*

the dams in the early 20th century and today. Clearly, federal policies that mandate pollution clean-up, remediation, mining regulation and clean air and water requirements all had specific effects in their sectors, but they are also reflective of a larger trend in the direction of a cleaner environment and greater eye toward sustainability, though progress on the latter has been slow for many of those concerned.

The social and legal shifts also occurred against a backdrop of significant global economic change. The most obvious local effect of this change came about in terms of significant population loss. These factors alone can influence, change or even destroy large technical systems. This is what happened in Johnstown, Pennsylvania—the dam and reservoir system that we are viewing in this work is an LTS that was itself a part of a much larger system. With the extinction of the primary steel manufacturing industry in the area, much of the usage of the physical plant has downsized or extinguished,[8] but the remnants of some of the related systems are still in place. At the same time, there is an enormous public stake in making sure that those remnants—a system unto itself—continue to be viable and function well.

What makes the LTS construct an inviting and suitable framework for an exploration of the system around Johnstown is at least threefold: first,

Pittsburgh's Homestead works in 1907, showing the interdependence of steel, coal and rail. *Courtesy LOC.*

the interconnectedness of it, purpose-built toward the supply of a natural resource to an even larger industrial system;[9] second, the construction itself consisting of its physical attributes, the characteristics of its interconnectedness with the natural world since its inception; third, the most active question in this work, which is the conversion or transition of the system in alignment with larger social, political, economic and environmental changes and the criticality of its own infrastructure. This last point is further emphasized with the recognition of the singular importance—and silence—of maintenance of such a system and the challenges of doing so when the original builder of the system no longer exists. The system represents a kind of partnership—an uneasy one in many respects—between the built world and the natural one; the past, the present and the future; and the exchange between public and private goods and the thread of the continuity of meaning in the system, one of not only industrial heritage and legacy but also appreciation of the natural world and the ecosystem in which it has been situated. The area now has no primary steel industry, but it has what might be termed a "techno-natural" system.

Water has never been far from the common public discourse, at least in any given part of the United States during any given year. In more recent times, especially during the profound population growth in the American Southwest, water access and riparian rights are of top concern. Water for agriculture, for drinking and for recreation during a time of great population growth puts pressure on a limited resource. The enormous Oglala Aquifer, which supported the West for generations, is now largely gone. Dammed Western rivers run dry during drought. The situation has developed to the point where some serious proposals have been put forward, such as enormous pipelines from the Great Lakes to the Southwest. In the Eastern United States, the engagement with water has been far different; not nearly in such

Hinckston Run meeting a human-made wall, showing the techno-natural mix. *Author's collection.*

regular shortage as west of the Mississippi River, water's very omnipresence creates a different reality, particularly when it is an input to a larger system. It is always woven into national, regional and local politics.

The very idea of a "system of water" is sublime, prosaic and somewhat odd in any number of contexts. And yet, water is also something that is so frequently overlooked in the face of an ever-tighter, ever-maddening news cycle predicated on new technologies and the "cutting-edge." The essential function of water is that of life itself; without water, our lives are impossible.

The large technical system of the dams and reservoirs and industry around Johnstown made water into an instrumentality; the water was unquestionably used as a means of transportation, much like in other studies of such systems. It was also a catalyst for industrial transformation, a conveyance by which its own motion showed the manipulation of and control over nature, a natural system mixed with human intellectual and physical labor. That was the intended use of the system for decades, almost a century all told. The other aspect of the LTS and water is its "side" use, its adaptation as an area of outdoor recreation, which started not with the close of that original intent but an adaptation and usage that had been present from the very beginning of the system and ran parallel with the original intent. In addition, and as we will see in looking at the history of the system that is the subject of this work, the flexibility and adaptation of usage by different social groups were clear at the outset of the creation of the system, even as it was heralded as an example of engineering wonderment in the midst of industrial might.[10] In essence, while the system was supplying tens of millions of gallons of water per day to make steel, people also started to go fishing. We bear this in special mind as well when we note that the fish didn't just magically appear in the brand-new lakes. They were *placed* there.

This book is a history of the dams, to be sure, but also a history of heavy industry in the region; its role in the history of the country; and the people who have lived and worked in, around and subject to this technical system of railroads, coal, steel and water. This is a story told from the standpoint of the seldom-recognized inputs to the industry: the natural world and the infrastructure that, once built, disappears from conscious sight and seldom reappears except for instances of emergency, fear or one of enormous transition. This book discusses all three of these factors, but the one that is of the greatest interest here is that of the latter: how an LTS transforms into a new purpose—or, indeed, to question whether it even truly does. It is, after everything else, a history of the people and the place.

We therefore see a push and pull, interwoven among many layers. It is a story of significant color and interlinked factors that defy easy categorization. This is not just a chronology of technology; this is a mosaic of people and things, technology and change, the built world and the natural world. It is best to bear this in mind as we explore the narrative of these considerable but hidden dams and the natural world's wonders that manifest from human hands from barely a century before.

A Word About Maintenance

Part and parcel of a large technical system is its maintenance. For too many this seems to go without saying; for without saying it, it then goes without consideration and then, of course, budgeting. One of the elements of maintenance in this history, steeped in the reality of Johnstown and heavy industry, is the *lack* of maintenance—or even the opposite of it—that occurred in the wake of the damage and death sewn by the failure of the South Fork Dam and the great flood of 1889. The concern surfaced in 1936; it surfaced in 2021. It surfaces every day for those charged with making sure the dams stay intact and the billions of gallons of water in the reservoirs remain penned behind them.

One of the other aspects of a study such as this is something that seems to occur unless and until a political debate sparks sudden interest, and that is the phenomenon of the "disappearing infrastructure." At some point in the individual and shared human experiences, most of the users of an infrastructure system forget it is there. It is, in essence, taken either for granted or as a given. It is only with the reminder of one's attempt to flip a light switch during a power outage or the surge of panic that occurs with a broken pipe. That's at the level of a single household.

When one considers something like the highway system—that is, the system that has been extant for years or even decades, rather than the perpetual "new" construction that seems to be favored—this invisibility of the infrastructure occurs at a broader scale, almost that of a mass delusion. In the saga contained in this work, the disappearing infrastructure reappears at times of either political or environmental crisis.[11]

There has been recent and increasing political rhetoric and also real policy proposals for the proper deployment of maintenance in the United States; often, it comes about in the context of large-scale infrastructure bills, but as Vinsel and Russell note, inclusion of infrastructure in such enormous government spending is usually intended in terms of construction of *new* roads—roads that will, one day soon, also require maintenance. "Despite the importance of maintenance, we often fall asleep on the job. Businesses, homeowners, governments, and other groups responsible for public infrastructure often respond to the high costs of maintenance by ignoring them. Deferred maintenance can impose significant and even tragic costs."[12]

There may be no truer proof of that statement than the truth that the people in this region, and really all of Appalachia, know well—falling asleep at that job can cost lives, perhaps catastrophic loss counted in the thousands. Deferred maintenance on a dam impounding eleven billion gallons of water is the cognitive equivalent of a death sentence for the people whose only "wrongdoing" was living in a spot downriver. In short, there is no real room for maintenance deferred.

Vinsel and Russell's main thrust is an argument against glib innovation-centric buzzwords, but some of the other points they make are apropos, particularly in the context of the LTS, power (see below) and the realities of historical events in the area. "Much of human history is, in fact, stories of stability…Technology, as a human phenomenon, is much broader and deeper [than new information and computer 'tech']—it encompasses all of the materials and techniques contrived by human civilization, including non-digital technologies such as guns, sidewalks, and wheelchairs."[13]

Another set of theories, besides that of the LTS, is the product of the fertile mind of historian and philosopher Michel Foucault, and it is worth brief consideration of his thinking about the connotations meant by the word *power*. Foucault was fascinated by conceptions and conflicts of power and discipline, although his full intentions with those terms were rather technical, highly specific definitions that elude the usage that we deploy every day.

Notions of physical power: pushing, pulling, along with its ebb and flow, its sharing, its passing along from one side to another, carry forward throughout many of the elements present in this work. The power comes to the conversation in many different directions; certainly, economic power is one of the areas that should be borne in mind, particularly when one thinks about the resources spent by Cambria Steel and its successors in the interest for the construction and upkeep of the dams and reservoirs. Parallel with this is the competing idea of power between the original landowners of the areas that Cambria Steel had in mind for reorienting use as a water supply. Tense negotiations and litigation were common occurrences, the legal system used as a dispute resolution machine to head off what could have likely been violent confrontation in a different society. The major labor movements that eventually gained greater traction later in the 20th century are fairly well-documented, for example, but the hills of Somerset County were quite restive in the days and years leading up to the eventual construction of the dam and reservoir. The power exhibited in these circumstances is much the same as we might witness today, though with likely greater secrecy and more violence. An instance in 1904 in Boswell—close to the eventual site of the Quemahoning reservoir—was of the violent suppression of a "riotous" strike of coal miners who walked off their jobs without notice in support of other miners in the nearby communities of Meyersdale and Garrett. The intertwined relationships and push and pull of power have long been in place in the area.[14] The news reports of armed farmers standing athwart steel company men were just other examples of a restive area in a difficult time of accelerated social, political and economic change.

Different paradigms of power developed and took shape as the steel industry faced significant contraction and Bethlehem Steel slowly staggered toward its demise. The wrenching effects of these changes brought significant upheaval and discord to much of the social fabric of the area, with population losses exceeding 50 percent per decade in the city of Johnstown, with younger generations searching for opportunity elsewhere. Power comes to the forefront even more in the realm of local politics of public and private ownership of the water resources—more lawsuits, certainly, but also electioneering and showing a profound and significant shift toward seeking public funding as opposed to private sector growth.

Power takes many forms, from the smallest hint of influence between two individuals to changes in vast populations in the region, the country and the world. Johnstown and the area at large have been participants, sometimes consciously and sometimes not, in all of those. The effects remain, and the changes continue.

Bethlehem Steel in Johnstown

This section explores the history of Bethlehem Steel in a general sense and the company's relationship to the Johnstown area. The cultural connotations and connections during the mid-20th century represent an important dimension to placement in a broader historical context for the community and the firm itself. Indeed, though Bethlehem was not the largest of the steel producers, it was still a colossus; the influence it had over Johnstown and its surrounding community is hard to overstate. Similarly, the impact of its collapse would be, quite simply, catastrophic, far more catastrophic in an economic and social sense than the flood of 1889.

It may not be out of place to give several personal recollections on this, as Bethlehem as an entity had its final collapse in the year of my high school graduation, 1992, though it had mostly evaporated from Johnstown ten years prior. The Johnstown School District's junior high school was, at the time I attended, located in the west end of the city (a major part of which is not coincidentally named Morrellville, so named after Daniel Morrell, the former managing director of Cambria Steel in the 1860s and 1870s). The bus route took me past the sprawling layout of the Lower Works stationed along the Conemaugh River, where my grandfather once worked, starting at the Stone Bridge and the Point where the Little Conemaugh and Stonycreek combine. Ancillary operations for the Lower Works stretched downriver with power plants and steam turbines, paralleling Broad Street into the West End of the city. Broad Street ends at the entry gates of what was then another Bethlehem facility, the Wire Mill, thus capping the end-to-end nature of the sprawling and dominant Bethlehem facilities in Johnstown.

One of the romanticized, laissez-faire fantasy depictions of the late 19th and early 20th centuries was that of unfettered industrial and technological progress during an age of great economic expansion and opportunity. The belief holds that these "good old days" were ones of unrestricted business opportunity and a federal government that saw fit to resist intervention in the economy and allow the market to shape itself and, in so doing, shape the world and society as well. The vision of heroic captains of industry pulling themselves up by their own bootstraps to create vast wealth out of nothingness lingers well on to today, oftentimes predicated on a belief that a return to glory would surely follow a political and economic environment that would reflect the restoration of this lost golden age. We bear witness to constant repeats of this refrain in virtually every political campaign for virtually every public office.

Of course, many of the companion depictions showcase abject misery. The huddled masses, as hailed by the salutation at the foot of the Statue of Liberty,[15] remained huddled in vast squalor in poor, ramshackle tenement conditions. They endured punishing work schedules and work environments, in a new land rife with racism, ethnic hatred and discrimination of all shapes and sizes. Many lost their birth names upon entry to the United States.

Pictures of the working classes from this time span the gamut from grimy, dusty and desperately ill to those who are better dressed, better fed and in better health—likely far better than what they could have accomplished back in the "Old Country."

Both of these narratives could do with a bit of nuance and deeper understanding. Living standards did improve, but the struggle to accomplish that was sometimes difficult, violent and deadly, shown clearly by major events in the history of the labor movement, some of which are recounted herein. At the same time, some individuals clearly rose to the top and commanded incomprehensibly vast sums of wealth and power, driven in no small part by ambition and talent—that much is true, but it is far from the whole story. All these aspects had both upsides and downsides, and the resulting picture, stripping away today's politics and the layers of ideological rhetoric, is complicated.

What is undeniable, however, is that this mix of labor and capital, immigration and class status, created a political and economic web that is far more complex than most narratives, particularly those with political axes to grind or those seeking to impose present-day values and conditions on those of the past. One of the most important, if not absolutely most important, individuals in the corporate history of Cambria Steel was Daniel Morrell, a Quaker originally from Maine, who, along with some relatives, managed a company with a dry goods store in Philadelphia and the namesake of the neighborhood previously mentioned. This same company owned Cambria Iron, and Morrell came to Johnstown to manage it—indeed, much of his work was to save the plant from mismanagement, constant economic danger and an almost cartoonishly bizarre history of corporate fumbling. Morrell brought Cambria Iron back from the brink of bankruptcy to become one of the largest iron producers in the world in the 1870s and 1880s.

Morrell also had a stint as a U.S. congressman from 1867 to 1871, where most of his activity seemed to focus on tariffs targeting imports of British iron. The revolving door between industry and government is not a recent invention; one can see this clearly in Morrell's own political and business pursuits at the time. This is not a critique of his character but a mention of an important fact that better illustrates the reality that strays from the preferred dream of the laissez-faire champion. Morrell was also, in fact, far from the only one, and the great iron and steel concerns frequently cast long shadows over legislation and policy debates in Washington, D.C.[16]

Throughout the 19th century, the American economy followed a close series of booms and busts in a roller-coaster pattern more precipitous and volatile than witnessed over the 20th century. There were more than twenty depressions and recessions between 1830 and 1910, some of which showed a significant decline in business activity by greater than 30 percent.[17] This, coupled with the Civil War, numerous extraterritorial interventions, a plutocracy that captured a large portion of the federal policymaking system, significant labor, communist and anarchic agitation and populist upswells, helped create new pathways that included a significant new interventionist regulation in the economy. The results included strong new competition laws, safety regulations for food and the Federal Reserve system. We land now in the era of the early 20th century, the time when the dams were built and the reservoirs filled.

It is with these events in mind that we can consider the construction of the Quemahoning Dam and Reservoir at the close of the first decade of the 20th century, a time of great industry consolidation and a young and growing country not only maturing but also adding to its economic strength, its international interventionism and complexification of its social fabric. The entrance of the large technical system consolidated during this period, ranging from vast systems of electrification and public works, to major industrial and financial conglomerates formed by the likes of JP Morgan. The latter is the party responsible for the formation of the massive United States Steel (USX), the dominant player in the steel industry for most of the history of the period of industrialization in the United States.

With a colossus such as USX looming—sometimes figuratively, sometimes literally—over the landscape, the most natural business response for a competitor is to get larger as well. Bethlehem Steel, headquartered in the eastern part of Pennsylvania, sought growth too, in an attempt to keep up with its larger rival. Increasing the size of a firm frequently drives costs down, increases bargaining power, creates more pathways to produce and raise capital and enhances an entity's political standing and ability to influence public affairs. The momentum gathered by these maneuvers oftentimes generates its own further momentum, until one reaches a point of slowing growth, maturity or perhaps even decline. In terms of the legal, statutory and policy changes wrought by the Gilded Age, with the establishment of the Federal Reserve and new governmental enforcement mechanisms against large trusts and anticompetitive rackets, the business environment for heavy industry was somewhat more restrained, but the overall economic trends toward growth and its necessity to help ensure a firm's competitiveness were still intact.

Just before 1920, Cambria Steel merged with Midvale Steel, a company that had also started in the Philadelphia area. Midvale specialized in ordnance and armor production in its manufacturing lines, and the acquisition of facilities such as those at Johnstown pushed in the direction of defense spending and armaments (also, notably, a swath of customers consisting of government spending, not the least of which was the United States). In just a few years, however, Bethlehem swallowed up the whole of Midvale Steel and all of its facilities, including that of Johnstown. These events coincided with the peak of the steel industry in Johnstown, along with its demographics. Its peak population neared 70,000 in 1920 and has steadily lost population since then, to its current level of just over 18,000.[18] The footprint of the steel industry in Johnstown consisted of several major facilities: Gautier, the Lower Works, Franklin and the Wire Mill. The mills alone employed tens of thousands, with ancillary enterprise—ranging from coal mines to restaurants, contract partners with the mills, restaurants, shops, inns and the rest—tens of thousands more.

The years of greatest interest in the first regard are those prior to the events leading to the United States' direct involvement in the First World War; the easiest way of bracketing the dates would be approximately 1899–1915. The line of presidents during that time included McKinley, Teddy Roosevelt, Taft and Woodrow Wilson. The times were that of significant social, political, economic and demographic change within the United States and certainly magnified in western Pennsylvania with waves of thousands emigrating from Europe to work in the mills and mines. Conventional wisdom might suggest that the country's entry into World War I was the point at which the United States became an international

player and signified its loss of isolation (and innocence), but this is not truly accurate. During this time, the United States undertook a period of economic and population growth so significant that its international effects were profound on the basis of its burgeoning GDP alone, with massive population fleeing Europe and parts of Asia to a strange new country and forming a warp and weft of greed, political activism, profound racism and integration and mixture of new languages, cultures and understandings—and also lack of understanding.

By 1915, Los Angeles already had around a half million people living there, as did San Francisco, despite the great earthquake and fire a few years before. Even Seattle had more than 200,000 inhabitants by then; the railroads had fully tied the vast land area of the United States together. Cambria Iron and then Cambria Steel had played an outsized role in this expansion; since the 1860s and 1870s, it was the leading producer of railroad rails. Decade by decade, transcontinental travel had been established; one could cross the country in three days just a few years after the Civil War. By 1915, such feats were old news and about as remarkable as any of the dozens of transcontinental flights that thousands of passengers will participate on any given day today. The railroads were the customers that truly placed Johnstown on the map.

During the second decade of the 20th century, the United States was a rumbling and growing, jagged and uneven social and economic mishmash, a surging agglomeration of population and monetary growth. The interplay between the politics of the new immigrant arrivals to the country along

The sprawling Franklin works of Cambria Steel, circa 1900. *Courtesy LOC.*

Blast furnaces 5 and 6 and hot blast stoves of the Lower Works/Cambria facility of Bethlehem Steel in Johnstown, mid-20th century. Constructed circa 1876, demolished and scrapped, 1986. *Courtesy LOC.*

The Gautier division of the Bethlehem facilities in Johnstown. *Courtesy LOC.*

with the generations that had been in place and were now largely in positions of power at every level of government was a significant one. Even with the friction this would cause—including some displacement of the African American population uprooting from an oppressive South and to a marginally less oppressive North—there was no true reason to believe that growth would reach any type of hard stop.

The steel industry, and Bethlehem along with it, continued to be an industrial segment of great economic importance to the nation. Struggles with labor unions were not infrequent during these times for any of the large heavy industries. In nearby Windber,[19] Pennsylvania, major labor strikes involving coal miners occurred in 1906 and 1922–23, bringing not only state government but also national attention to the situation between workers and management-ownership. Big steel had a broad menu of troubles as well, including violent uprisings, such as the famous 1892 Homestead strike near Pittsburgh, all the way through the interwar years. A national steel strike in 1919 deeply influenced Johnstown labor, just as it did in many parts of the country. More than 30 years later, in 1952, President Truman essentially nationalized the steel industry to quell a labor dispute, though the Supreme Court ruled that he exceeded his authority in doing so. That said, the effect of hundreds of thousands of workers and their families pushing, agitating and fighting for a greater share of the economic bounty over the course of several decades was profound and the actions prominent in the national consciousness.

Production and disputes all had costs and as such could be accounted for. Later, the toll on natural resources would also enter the ledger. Before that point, however, it still required accounting as an input and that input would often be significant. With the Johnstown works in operation, it would consume one hundred million gallons of water a *day*. No water, no steel.

As far as local effects, Bethlehem Steel in Johnstown continued the long-standing actions initiated by Cambria Iron in the latter part of the 19th century by providing amenities to its workers and to the people of the Johnstown area. In earlier times, Cambria Iron, under the leadership of David Morrell, established a community library, a water system (predating the Manufacturers Water Company it incorporated for the construction of the Quemahoning), a hospital and other town developments. We enter the story proper of the dams in the early 20th century, just before the First World War and before Prohibition. We will see how the LTS changes and evolves and, as decades fade, Bethlehem Steel's actions regarding production, wartime and organized labor. The efforts were sometimes absurd and at times inhumane but also included long-running community support and many amenities for its workers. One was what was known as the Management Club, a large Tudor building serving meals overlooking a golf course and tennis courts for members of the management team in Johnstown and those visiting from other sites. The other was called Bethco Pines, a camp and resort area on the shores of the Quemahoning Lake, the crown jewel of the system.

2

THE BEGINNINGS OF THE SYSTEM

1900–1936

June 29, 1900: The heaviest rainfall of the present season is reported from Quemahoning Township, Somerset county. About 6 o'clock Sunday evening, says the Somerset Herald, *rain began to fall in torrents and in a few minutes the water in Lafferty's run* [sic] *had swollen way beyond the banks. The old stone bridge near George W. Miller's place on the Stoyestown* [sic] *and Greensburg pike was practically ruined, and it is said it will cost $5000 to put it in its former condition. Old residents say that the water in Lafferty's run has not been as high since 1828. Fences were carried away, fruit and sugar trees were torn down by the storm, and growing corn and potatoes were ruined.*

June 29, 1900: Farmers of the Quemahoning Valley are armed and guarding a point that has been selected by the Cambria Steel Company for the erection of a dam [sic] *four miles long and having a depth of 75 feet at the breast. The farmers declare the dam will be a menace to public safety and decrease the value of their lands. The property owners below the proposed dam claim they will be in constant danger of a repetition of the Johnstown flood disaster. A clash between the civil engineers and the farmers is expected at any time.*

July 31, 1900: NEARBY TOWNS: Johnstown people threaten to dynamite the proposed Quemahoning dam.

August 10, 1900: Notice has been served this week on all the farmers in the valley who refused to sell their land to the Manufacturers Water Company, an auxiliary of the Cambria Steel Company, that condemnation proceedings to gain possession of the land are to be brought. Citizens of Johnstown say they believe that if the dam is built it will not be permitted to stand a single day unless guarded at all times.

Building the Quemahoning Dam was not an overnight project in terms of either its construction or the planning and preparation. At the time of its inception, it was well-known and understood that this would be the largest lake in the state of Pennsylvania, and Cambria Steel was building it within the living memory of many of those who had survived and witnessed the horrors of the flood of 1889.

By 1917, the United States was already a sprawling country spanning a continent. Even with burgeoning air and automobile transport, for example, the railroads were still king, and railroads always needed steel—for rails, for locomotives and for cargo and passenger cars. Railroads still needed coal, and steel mills still needed coal to make coke for the furnaces. Leading up to this in the first years of the 20th century, Cambria Steel was a complete and integrated steel plant manufacturing everything from shovels to wire to milled steel ingots. For decades it had been a leader in railroad rails, and the industrial growth of the United States, even with its jarring boom-and-bust cycles, was unrelenting.

Thus, the enterprise needed more water and a dependable source at that, preferably within the control of the mills themselves, as their economic goals had to take pride of place in their considerations of an environment with heavy competition between many large rivals. The prominence of the Quemahoning Dam was virtually guaranteed, with its association with a massive set of steel plants, the fresh memory of a major cataclysm a few years prior and the crop of superlatives wrought by any large construction project even today. This created a systematic and thorough record on which the historian and enthusiast can draw, as the process of the build was well documented through photographs, records and publications.

John Robson, quoting George Donehoo's research from early in the 20th century, stated that the name Quemahoning stems from Cuwei-mahoni, or "Pine Tree Lick."[20] Robson searched for the place that any such Indian village may have been located but did not meet with any success. There was a village nearby, Kickenapaulin, which does have historical attestation and was located in an area that would be consistent with the record and with physical evidence, between Bedford and Pittsburgh, along what was called the Raystown Path. Robson, again quoting Donehoo, noted that the trail, leading between the Susquehanna and Ohio Rivers, was very well-known and highly used, handily predating European colonization. Robson noted his efforts to locate the village but seems to be inclined to believe that the any village likely to be Quemahoning was, in fact, Kickenapaulin.[21] He also noted that the artifacts with which he was familiar in the area indicate quite old dates, back to the Archaic Period of the eastern woodlands, meaning that they date back to at least 1000 BCE.

Somerset County has the highest average elevation in the state, along with its tallest point (Mount Davis) and some of the most rugged overall terrain. Long a period of frontier life from the earliest times of colonization, Somerset County is still largely rural today. Iron manufacturing pulled westward from the earliest plantations in the eastern part of Pennsylvania, between Philadelphia and Reading, and converged in spots such as Pittsburgh and Johnstown. Even so, one of the earliest "iron plantations" in the state is in northern Somerset County and dates to the early 1800s. Somerset County's most significant industrial-age contributions, however, came from the coal industry, and the county possesses many deep and surface mines alike. The aspects of water transport, mineral rights and ownership and agricultural land interests all combined to create the story of the Quemahoning Dam and reservoir.

The initial water supply from the Cambria facilities was built in 1867. Given the growth of output over the next forty years, the need for additional water grew more pressing. Steel making takes not just ore and heat but enormous quantities of water. A mill that produced 25,000 tons of steel in 1872 would, by 1912, produce 1.3 *million* tons. At the same time, the local population of Johnstown grew from 6,000 to 60,000. Combining these two measurements, steel grew by 5100 percent and population by 900 percent. It is not just the internet age that saw the famed and highly coveted "hockey-stick" growth rate.

Cambria Steel formed a subsidiary, the Manufacturers' Water Company (MWC), in February 1900, chartering and charging it with the goal of finding alternative water supplies to anchor the existing operations and to enable further growth. In that same year, MWC began to conduct surveys in the area while those in the mothership of Cambria itself created new plans to estimate the expected growth over the next decade or more—along with protests and threats of violence.

One of the central considerations in the analysis of a large technical system is the inclusion of knowledge, planning, financial controls and other social measures in the analysis, along with an examination of their enrollment in the technical construction process itself. An article from the *Engineering Record* journal is important in its early recognition of this, noting throughout its recounting of the construction of the dam and its history, those areas where the business planning and decision-making aspect occurred in conjunction with the engineering choices. The other aspects of the social considerations involve those citizens affected by the construction of the dam, either in immediate proximity to it or more distally, and many of those aspects are considered here.

Understanding the actual construction of the dam is an important consideration, not just at its time, or its time at the present, but all through the span of decades. It is still notable in the writings and depictions during its planning stages, especially its anticipated scale and the goals of its operation. The surveys that MWC undertook in the early 1900s allowed the planners to note that the area of the Quemahoning Creek, a tributary of the Stonycreek River, would work to be the new centrally important water supply for the entire footprint of the Cambria Steel enterprise.

The area of the Que surveyed included 92 square miles of water catchment and drainage, which was calculated to allow for a daily usage of 90 million gallons of water,[22] enabled by the construction of a dam wall nearly 100 feet tall from the valley floor. The type of construction of the dam took on even greater importance not just in the face of ordinary public concern but with the understanding of the regional cultural fear surrounding large dams generally and the specific events of 1889.

MWC dispatched envoys on a scouting mission to the western United States and Mexico to investigate dam construction and types and bring back learnings relevant to Somerset County conditions. After a thorough study of eight different dams, the consultants returned a general outline for the specifications and requirements for an earthen dam, which, of course, did little to allay the social concerns of the time. Given the political and economic powers of any large manufacturing firm of the era, and especially the primacy of Cambria Steel in the Johnstown-Somerset region, opposition would be overcome.[23] Nonetheless, the hydraulic-fill dams throughout the southwestern United States created solid models and precedent for the construction of this new reservoir.

The geography of the area in this new construction came to the assistance of its build rather than providing the pathway toward mass destruction, as it was commonly feared in some quarters. The valley walls were quite high and steep around Quemahoning Creek, with a soil composition favoring heavy clay. The shale rock and clay were excellent materials for the formation of such a dam, and both were in significant abundance from the hillsides themselves.

The methods of moving the materials for the construction also used nature; local timber would be felled to create great sluicing conduits with the abundant water to spray and use hydraulic force to move the clay and shale rock into position in the area surveyed for the breast of the dam. The MWC personnel also made sure to prototype the dam at reduced scale, and in so doing, their hypothesis regarding the sturdiness and durability of the clay and shale dam was proved out.

Technical Features of the Quemahoning

While the surveys were conducted in 1900, construction didn't begin in earnest until 1907; there are any number of reasons for this, but some of the more specific and local interests are worthy of mention. The size of the undertaking was significant, just at the outset; the MWC and Cambria knew this would become the largest body of water in the Commonwealth. The MWC also initiated legal action as early as 1905 in an effort to acquire the land in the Quemahoning area; estimates at that time were for 50,000 gallons per day. Of course, this was a significant underestimate of what would eventually have occurred, whether that was the early target or something that would sound more palatable than the millions it would become. Early news reports also noted, in 1905, that the planned body of water would cover current farmlands. It seems reasonable to infer that many of the area residents, even years prior to the actual commencement of construction, knew of the plans to build the dam and its rough dimensions. There had been significant, and deeply concerned, communication among neighbors.

Given the concerns regarding the possibility of a dam, there were plenty of those downstream of any dam failure that registered steep concern, including many inhabitants of Johnstown, 14 miles from the construction site. These can rather easily be understood and grouped together, as they were voices raised with worry with the none-too-faded memories of 1889. There is likely to be little question that many within the ranks of Cambria Steel and MWC also had the same worries; the construction of the dam, as it turned out, would be at least somewhat guided by an abundance of caution and reinforcement in the components and methods.

Once the paths were finally clear enough for MWC to move forward, however, the largest lake in Pennsylvania did not take too long to appear. The flumes, made of maple wood and reinforced with steel plating and used for the construction and hydraulic movement of the rock and fill, followed a plan of deposit, with two 1,000-horsepower pumps pushing enough pressure to reach 75–110 pounds per square inch, with the fill and the rocks moving along the flumes to be deposited with the gravity, falling or filtered from the flume at different points. The capacity of the pumps each could push 5,500 gallons per minute, while both pumps used together would combine that capacity to total 11,000 gallons per minute. All the machinery and labor benefited from the steepness of the valley walls, which made the construction dam both closer and assisted by gravity; it was obviously more of a matter than things simply falling into place, but the engineering solution was formed and even inspired by the topography itself to create a structure of robust composition, the techno-natural system taking its incipient form. This was all done with the social and cultural background of the fear of inundation and even total destruction.

The dam would end up being nearly 1,000 feet long and almost 100 feet high; as placed where it was sited, it was located 14 miles south of the city, at an elevation of 1,540 feet, which would be in the range of 200–300 feet higher than the three major steelmaking plants the dam would supply. The spillway sat at 1,620 feet above sea level and the crest of the dam itself at

This page and opposite: Construction of the Quemahoning Dam, 1909–11. *Courtesy CSA.*

40

1856-4-3-11

1,633 feet. The steel pipe, initiated at the gatehouse of the dam, ranged from 66 inches to 58 inches wide, manufactured from riveted steel plate.

The lake would hold 11 billion gallons of water,[24] with the possibility of 12.3 billion gallons with the spillway flashboards placed; this volume of water with the 66-inch pipe generated 190 pounds per square inch of pressure, which would be reduced to around 50–60 pounds per square inch by the time it reached its outlets at the mills miles downstream. The width of the dam is 20 feet at the crest and 700 feet at the base, which was dug all the way to reach the bedrock with a concrete core wall to prevent seepage. The aforementioned spillways ended up being 250 feet wide.

The dam itself used 600,000 cubic yards of material, and the reservoir formed behind it ended up with a length of 2.5 miles and a width of a half mile, covering 900 acres of land with water. The concrete outlet of the dam, which included the riveted steel pipe, was 30 feet wide and nearly 22 feet tall. Construction took two years and two months to finish, although four of those months were spent at idle due to heavy weather; Somerset County is still notable even today as among the coldest and snowiest counties in Pennsylvania. Builders took core samples of the dam and found that the structure's constituent materials had solidified into the

This page and opposite: Construction of the Quemahoning Dam, 1909–11. *Courtesy CSA.*

anticipated mixture, confirming their confidence in the design and build of the shale-and-clay work.

The main pipeline leading from the dam itself is the next major feature of the engineering feat. The pipe itself conveyed water 14 miles, starting at 68 inches in diameter at the gate house of the dam, reduced to 58 inches at the lower end, splitting into three separate pipelines—one of each of the major facilities planned to receive water from the dam—at the edge of the city. The water flow is governed largely by gravity and pressure and, as mentioned above, is made of riveted steel supplied by Cambria Steel itself. Many areas of the pipeline are cradled and covered in concrete for reinforcement purposes, particularly where the pipeline had to cross the winding course of the Stonycreek River. The trenches for the pipe were about 7 feet wide; the trenches varied from 7 to 14 feet deep.

While the course of the pipeline occupied 14 miles, the pipeline itself took 73,100 feet of pipe, which is just over 12.5 miles. This included 9,000 combined feet of four tunnels carved at times through solid rock, with a bore dimension of 9 feet by 9 feet. The steel plate used to make the pipe ranged from 5⁄16" to 5⁄8" in thickness, more than enough to contain the maximum calculated pressure of 185 pounds per square in (psi). Controlling the flow along the length of the pipeline are a number of blow-off valves, backpressure valves and 36-inch gate valves.

Disputes, Acquisitions and Settlements

We can now revisit some of the news pieces that opened this chapter: threats of armed standoffs and violence. While there is some conflict about the actual *beginning* of construction at the Quemahoning site, it is clear that Cambria Steel had been eyeing greater water access for quite some time prior to construction. In 1900, for example, there were reports of armed standoffs occurring between the farmers and landowners in the area coming into contact with survey crews from the Manufacturers Water Company:

> *Farmers of the Quemahoning Valley are armed and guarding a point that has been selected by the Cambria Steel Co. for the erection of a dam four miles long* [sic], *and having depth of seventy-five feet at the breast.…The farmers declare that the dam will be a menace to public safety, and decrease the value of their lands.…The property owners below the proposed dam*

> *claim they will be in constant danger of repetition of the Johnstown flood disaster. A hand-to-hand battle between the civil engineers and the farmers is expected at any time.*[25]

Notably, I've been unable to recover any records of the results of a fistfight (or more) between farmers and engineers in the area circa 1900. Notwithstanding the rather breathless final sentence, this is an early reference to the tensions that would have invariably developed given such an engineering project. In fact, a few weeks later, there are reports that the survey crews gave up in the course of the confrontations with the landowners in the area. Again, the reason listed in the news reports is fear of a replay of the 1889 catastrophe. Other sources note, briefly, the possibility of a case of engineering wonder, as in a brief report from Indiana, Pennsylvania, stating that the project will provide employment for 400 men and a "year's time" for the construction of the project (which, as we will see, was an ambitious target).

One might be forgiven for expressing some degree of skepticism regarding the claims of fear, danger or comparative safety from the dam, especially once Cambria Steel, through its MWC subsidiary, began to open its checkbook and make things happen. What was stopped with threats of physical violence, the demonstration of firearms and blocked access to survey crews was easily swiped away by the presence of pen and ink and a generosity with the number of zeroes written on a check.

The borough of Holsopple is located along the route of the main pipeline between the reservoir and the Cambria Steel facilities. The main ran through the borough and served as the main water supply for firefighting purposes, with four hydrants included in the improvements in the town infrastructure around 1910. In addition, a later retrospective of Hollsopple stated that it owed its existence to a railroad being built between Somerset and Johnstown that later became part of the larger Baltimore and Ohio Railroad Company; these tracks were also heavily used by Cambria Steel in the first decades of the 20th century for the logistics and transport of materials and resources for the construction of the Quemahoning Dam.[26] Many regular news reports concerning the construction of enormous railroad tunnels were announced, along with surveys and estimates of the findings of enormous coal fields in western Maryland and southwest Pennsylvania. News reports of the "huge" dam to be constructed in the area continued to make the news in regional and national newspapers for years, even before construction truly began.

Besides all of this, there is documentary evidence suggesting that the MWC's movements to begin the Quemahoning construction span back several years before the project first began to make the rounds of newspapers.[27] Rumors undoubtedly started early, and even in the days before the internet and even truly before widespread usage of the telephone, the fence-line gossip was an effective and efficient way of transmitting information from farmer to farmer. By then, it was not only the farmers of the Quemahoning Creek area but the people downstream as well who would threaten dynamite to any construction.

The dam was never dynamited, of course, although threats of doing so, especially during wartime in later decades, were certainly present. This emphasizes the roots of power and its dynamics in historical, economic and social terms in the region. It also shows that there was a transition from the rugged colonial "self-help" type of power to that of dispute resolution in a more modern sense: lawsuits. Indeed, there was some litigation involved, but mostly along the lines of the brinksmanship that it typically takes to arrive at a settlement.[28] One notable landholder, Joseph Rininger (described as a "modest farmer of Somerset County" by the Lancaster *Intelligencer* in February 1907) received and cashed a check for $200,000.[29] This is roughly equivalent to a payout of $6.5 million in 2023 dollars; this "modest farmer" became a wealthy man. It wasn't an overnight success—after all, he held out against offers and imprecations for around six years before he received this check. All of that aside, however, it was certainly life-changing for Rininger, who had purchased his 400-acre farm eight years prior. It appears that Rininger himself had some steel in his bones, as described by the *Intelligencer* in 1907, in recounting the negotiation after MWC had surveyed and purchased many, if not all, of the other surrounding properties:

> *Rininger was asked to name his price. After due deliberation, he said he would take $100 an acre. The figure quoted the agents said was preposterous and the left the valley, obviously believing that Rininger would soon accept a lower figure. Four months later they returned and learned that he had raised the price, and when they (MWC) made the third effort to purchase the farm he laughed at their proffer of $100,000. For several years the purchase of the Rininger tract was lost sight of, and meanwhile a rumor was industriously circulated that the Cambria company had abandoned the idea of turning the valley into a water basin. Rininger smiled good-naturedly when his neighbors twitted him for letting*

> *a golden opportunity slip through his fingers, and patiently bided the time when he could in turn have the laugh on them. Friday he startled the whole countryside by exhibiting a check for $200,000 drawn on the Girard Trust company of Philadelphia.*

This occurred after several accepted offers that the newspaper itself described as "ridiculously low" prices. Rininger would likely have made an excellent poker player.[30] Other reports indicated that that the project to build the dam was going to move forward immediately after a few final pieces in the property puzzle that MWC consolidated to begin construction of the dam and reservoir proper, behind a dam breast "100 feet high." And despite—or perhaps because of—the closure of the deal between Rininger and the steel company, the wrangling and dealmaking for other properties in the area had still not been concluded. A man named Blough contested a number of the legal machinations that had been in place for the relocation of roads. The prices of land rose sharply after the Rininger negotiations had effectively raised the starting point of all the subsequent negotiations. There are reports of early sellers regretting their action on learning of the prices that land would eventually fetch: in some cases, nearly $20,000 per acre, which in 2023 would be nearly $650,000.[31] In an example—if only a rumor reported by the Meyersdale Republic—the influence of social factors can have on engineering projects, Cambria Steel changed its plans for the height of the dam breast in order to make the ensuing reservoir smaller. Therefore, the company was not compelled to continue purchasing negotiations for still more land. Given such economic realities, it is perhaps not surprising that there would be such engineering compromises. After all, the most winning formula tends to be the financial one.

In 1909, the dam and reservoir were still in an anticipatory state, despite the desires to conclude construction in under a year. A brief news readout from Harrisburg in February of that year captures a balance between seminatural wonderment, engineering fascination and clinching the story with what, for the time, would have been an eye-watering sum:

> *Lake Quemahoning, which is to be created by a big dam in Somerset county, is to be a new source of water supply for Johnstown and the Cambria Steel Company. This artificial reservoir, which, when the dam is completed, will be the biggest lake in Pennsylvania, will be located 15 miles from Johnstown. The water will be conveyed by gravity in a steel-riveted pipe 66 inches in diameter at the lake and gradually reduced to 56 inches at*

This page: Construction of the Quemahoning Dam, 1909–1911. *Courtesy CSA.*

> *Johnstown. The work will be carried on by the Manufacturers' Water Company, a subsidiary corporation of the Cambria Steel Company, and the outlay will be $4,000,000.*[32]

This prompts the question for the placement of this story, however brief. Perhaps it's the superlative nature of the story itself—this will be the largest lake in the entire state. With this is massive piping and an even more massive cost. Without lauding the operations of Cambria Steel outright, this small piece uses a tone of mild awe common to the time of describing industrial achievement. Another Altoona newspaper in 1909 declared that the Que would "solve for all time the question of water in Johnstown's greatest industry."

Still, the battles did not end, even over smaller parcels. In what might have been a bit of a backhanded compliment, the judge overseeing the case between MWC and landowner Blough decided against an injunction to stop MWC activities, as the work was proceeding so slowly, the immediate danger that an injunction was meant to halt was not present. The parties eventually settled that dispute as well, for undisclosed terms. Another report in 1910 noted that the expenditure on construction (well before completion) by MWC was a (curiously exact) figure of $1,749,727, or nearly $57 million in 2023 dollars.

The social and economic situation and conditions under which the Cambria Steel Company were fraught. The slowdown mentioned by the judge did not seem to be wrought by difficult environmental conditions or mistakes during construction; instead, there was a labor shortage, the announcement of which had plenty of Know-Nothing xenophobia to make a strange counterpoint to such a plea for help. From the *Altoona Tribune*:

> *There is an abundance of labor, more or less skilled, willing to take light jobs but the company needs men who have muscle and can stand hard work. Most of this labor goes to foreigners and there are more of the latter on the payroll than many of the officials would have if they could secure Americans. Six hundred men could find employment, it is said, on the big Quemahoning reservoir, in Somerset county* [sic], *and on the pipeline. The work is being held back by the scarcity of workmen.*

Once the battles *did* end, construction was afoot. Controversy may have continued, but the die was cast for what would indeed become—for a time—the largest lake in the Commonwealth. As soon as that became the case,

The enormity of the pipeline from the Quemahoning Dam is evident, circa 1910. *Courtesy CSA.*

despite its preeminence as an industrial water supply, the environmental conservation value of the reservoir was also as nearly immediate as could be expected. By 1911, there were routine shutdowns of the furnaces at the Cambria Steel plants to connect the facilities to the water pipeline. Other reports around the same time noticed the buoyant conditions under which the steel industry in Johnstown seemed to be functioning. Judges issued orders, largely on the side of the MWC, with regard to the litigation involving taking and repurposing of land. Still some even had spiritual costs, as noted from the *Williamsport Sun Gazette* in 1911:

> *On Sunday the last public service will be held in the Pine Grove meeting house of the Quemahoning congregation of the Brethren. For fifty-six years this place of worship has been used and is now to be abandoned because its site will be submerged by the construction of the large Quemahoning dam. The Sunday services will be singularly impressive.*[33]

Many decades later, in 1969, a historical retrospective in the *Somerset Daily American* noted that, in the early days of the settling of Somerset County, wool mills were the most prosperous sector. It goes on: "one was located in Quemahoning Township and probably was later known as Morgans Mill, now submerged beneath the Quemahoning dam [*sic*]." Time marches on, as does the progress of the construction; by early 1912, the MWC and Cambria had made the news in many areas of the country, noting that the anticipated finishing date of construction of the Que would be October of that year. "During the past few days fifteen miles of the pipe line was [*sic*] thoroughly tested."[34]

October came and went, however, and, like large projects tend to do, things can take longer than what had originally been anticipated, as this brief piece from the *Pittsburgh Press* in mid-November 1912 attests:

> *BIG RESERVOIR ABOUT COMPLETED . . . There remains but six feet of earth to be hauled to the top of the embankment. Several days ago, the big steam pumps which have been hurling four-inch streams of water against the earth and clay on the hillsides adjacent to the embankment were shut down for good. This slicing work carried over 500,000 cubic yards of earth and clay into the dam structure, which now approaches 80 feet in height. There is about 40 feet of water in the reservoir at the present time and it is expected that on Dec. 1 the waste gates will be closed and the reservoir filled to its capacity. It will take more than a month* [to fill the reservoir].

This same sentiment carried through another Pittsburgh paper of the same day, running the headline: "State's Biggest Dam Almost Ready." A more elaborate article ran at the same time in Altoona, noting some of the engineering superlatives of the construction of the dam, including the twenty thousand cubic yards of concrete at the dam core, included to bolster the strength of the dam and to avoid catastrophe such as what was witnessed in 1889. The litany of numbers—largely the same as in the Engineering Record piece—comes equipped with descriptors that include the listing of the "eminent engineers" responsible for the undertaking of such scale and complexity. The piece also concludes with the note of the loss of a town named Dibertsville[35] and the fact that a significant number of local landowners became quite wealthy as a result of the negotiations and litigation with MWC and Cambria Steel. The valedictory tone of the piece brings about an end for the first stage of the dam.

At the time of the dam's completion in 1913, a report from the field engineer from November of that year included a number of measurements and specifications of the system as it began operation. The report was based on a series of tests of the pipeline and its water flow to the various mills, stemming from that initial 66-inch pipe, reduced to 58 inches near the Franklin facility and a 50-inch line supplying the lower works and Gautier. The test counted a maximum flow rate of 113 million gallons of water per day, reduced and tested at 86.5 million gallons for Gautier and the Lower Works, with 26.5 million gallons heading to Franklin, as other water supplies were shut off during the test. The test was voluminous enough to lower the water level in the reservoir. The system was initially tested in December 1911, but the test in November 1913 was the first one of the system at full capacity. While the test was ongoing, all of the many meters in the system were checked every 15 minutes for the pressure within the pipeline—all 63,846 feet of it, 66, 58, 50 inches of steel pipeline, riveted in some places, double riveted in others and triple riveted in still others. The system carried ranges of pressure from well over 100 pounds per square inch to just under 20. In conclusion, the field engineer expressed a degree of disappointment at the readings in the system, speculating both about shortcomings in the test methodology itself and in the possible results, concluding that the initial velocity of the water flow was substandard.[36]

AN UNEASY HISTORY

The early history of both Hinckston and Wilmore reservoirs does not appear to be as fraught as the construction of the Quemahoning. Both smaller dams were even closer to populations and, in the case of Wilmore, located along the same river that caused the great flood of 1889. Both dams were also constructed earlier than the Que; the Wilmore Dam, however, was not originally a part of this system, being originally constructed by the Pennsylvania Railroad and operated by a different organization: Summit Water Supply Company. The location of the reservoir is along the top of the ridge between Cambria and Blair Counties, with Altoona just to the east and Johnstown to the south and west. Bethlehem purchased the Summit company in 1960 and took over the maintenance and safety of the dam. The deal was signed in Philadelphia, and the cost itself was $1, payable from Bethlehem Steel to the Pennsylvania Railroad.

The essential first steps with the Hinckston Run dam were, as always in those days, not to study safety ramifications per se but to begin making

This page: Winter reconstruction of Wilmore Dam, circa 1960. *Courtesy CSA.*

preparations for real estate deals and possible litigation and property condemnation proceedings. The earliest records of Hinckston reach back to the late 18th century, with original landowners in the area on deeds. One such notable name is John Barclay, a major landowner around the area with claims dating back to 1794. Some evidence indicates property ownership even prior to the War for Independence, dating to February 1775, several decades before even the founding of the city of Johnstown.

The naming of the Hinckston Run and the dam and reservoir is more colorful and perhaps tawdry than what might be assumed. A research paper by Robert E. Francis asserts that a man named John Hinkson gave his name to the creek that flowed into the Conemaugh, and this occurred during the 1770s. Spelling of many words, including proper names, was not set in those times, so it is perhaps unsurprising that one sees different combinations: Hinkson, Hinkston and Hinckston. Are they too similar to be different? Perhaps—but the John Hinkson of that era was the type of man that makes it difficult to appreciate; he, along with a friend, were accused of murder and fled arrest from royal magistrates and the governor of the time, as it was alleged, with no small amount of proof, that he shot and killed an Indian named Joseph Wipey who was in a canoe and fishing in the stream that now bears Hinkson's (alternative) name.

Originally from Ireland, Hinkson appeared to have been one of the tough but morally marginal players who inhabited the first frontier of the United States, the Appalachian Mountains, and assigned his name to this spot in Pennsylvania and a handful of others in Kentucky. He has been linked to several parcels of land in western and central Pennsylvania and for his participation in the extortionate Lord Dunmore's War, along with the American Revolution. He trafficked with several of the notable but equally morally gray men such as George Croghan and Simon Girty, participating in many exploratory and hunting expeditions as well as actions that seemed to be based partially in religious fervor and partially in racial hatred. He undoubtedly served with valor in the Revolution, and he undoubtedly killed a fishing Indian in cold blood. His legacy was, and still is, as convoluted and difficult as the Appalachian Mountains themselves, a microcosm of American history.

While the stream itself is likely affiliated with this man, the major landowner in the area prior to 1800, as previously mentioned, was John Barclay. The families in the area rose and fell as they always did in the first full century of American history, and by the late 19th century, one of the major landowners in the area was Benjamin Benshoff. Deeds and

Hinckston Run, upstream from the reservoir. *Author's collection.*

conveyances around 1906 indicate a purchase of large tracts of land along Hinckston Run. Other transactions with landowners in the area thicken the records, showing a sequence of interests from the initial purchases and condemnation proceedings by Cambria Steel and the MWC, with subsequent ownership interests via MWC to Bethlehem Steel and finally culminating in the millennial zeitgeist of the transformation into CSA, discussed thoroughly later. The documentation also reveals the records around Wilmore Reservoir and its transfer of ownership from Summit to MWC. A letter from a law firm in April 2004 confirmed this, particularly after the conclusion of a lawsuit brought by the Greater Johnstown Water Authority, which had been dismissed. The letter stepped through the timeline of the events related to the transfer of rights and obligations through the MWC to CSA in the form of a stock purchase agreement by the counties for the remnants of the company.

The interest in dam safety is, as we've seen, not new particularly among those living in or around the Johnstown area, and the maintenance and upkeep of the dams have been a priority and in the awareness of the general public for a long time. Wilmore Dam, while constructed by the Pennsylvania

A view of the southern end of the Hinckston Run reservoir near the breast of the dam, the spillway in view. *Author's collection.*

Railroad, is older than the Quemahoning, and in the year before the Que was finished (1912), there was news in the *Pittsburgh Post-Gazette* of an effort at strengthening the Wilmore dam after a flood scare.

> *As a matter of precaution the breast of the big dam at Wilmore is being reinforced with a seam of concrete two feet in depth across the top, while the middle of the breast is being lined to a depth of 12 to 15 feet. The dam is unusually strong, it is declared, and there is no foundation for the reports of its weakening. The Wilmore dam is about an eighth of a mile wide and* [the reservoir is] *about three miles long.*

The language in an earlier 1911 piece from the *Post-Gazette*, written shortly after the flood scare that had occurred, reported that the "city councils approved a report of an investigating committee to the effect that a break in the Wilmore dam was an absolutely [*sic*] impossibility, and that residents of the Conemaugh [river valley] never have been and never will be in danger from the dam." Trading in such absolutes is inadvisable, of course, but 22 years after the tragedy of the 1889 flood, fear and rumors were quite strong.

Early Outdoor Recreation

Importantly to the ongoing historical thread of Bethlehem Steel's assertion of property ownership was the notion of the exclusivity of access to the system. Posted and no-trespassing signs, along with the repeated conventional wisdom of generations, told of this. However, we can also see throughout the course of the 20th century that many people from different publics used the Que, Hinckston and Wilmore besides ranking officials and executives within Bethlehem. There are many early examples of this, and of course, because some of the evidence is published news and the activities involve fishing and hunting, there are unfortunate and sometimes tragic consequences of the action. Because it is news, an event is more likely to be reported if it results in injury; in the case of Wilmore, there is a 1923 report of a 14-year-old boy accidentally shooting his father in the head. The father had been fishing while the boy had been shooting birds (though August in Pennsylvania is not known to be a time for bird hunting, other than crows), and both were local to the area.

It goes even further than this as well, at least in some reportage devoted toward Hinckston. A report from the early summer of 1912 indicates that the "exclusivity" with which Cambria Steel held its property had been reversed, with the invitation of a number of outdoor clubs to pursue fishing, shooting and hunting activities around the reservoir. The clubs participating used a "special train" to move their members to participation in a type of fishing derby at Hinckston, as reported in a major Pittsburgh daily:

> *This was the official dedication and opening of the dam for public fishing. It is on the property of the Cambria Steel Company. On account of the recent rains it was necessary to open several valves in the reservoir to drain off the surplus water. As a result the bass did not respond to the coaxing bait of the fishermen during the morning. Only a few catches were made. In the afternoon more than 500 bass were caught. The largest was 19 inches. The program for tomorrow will include shooting for the United Sportsmen's championship, 90 targets and sweeps at 90 targets, open to all amateur members.*

Indeed, the momentum appeared to continue, at least judging from another report a few months later:

> *Now the Johnstown sportsmen want extensive game preserves and fishing streams of their own. There is much wild land in Cambria county, and*

> *they want to get a section of this, stock its forests with game and its streams with fish, and enjoy life. They will do it, too.... There is* [sic] *lots of fish in the trout stream in Cambria county and the Hinkston* [sic] *Run dam of the Cambria Steel Company has been thrown open to the* [sportsmen club] *members after being rigidly protected for seven years.*[37]

Reports of "big fish out of Hinckston" were fairly common, leading up to the public fishing derbies and shooting competitions. Indeed, one Daniel Berringer appeared to be particularly noted and, judging on his actions as reported in a Buffalo, New York newspaper in 1914, seems to have been party to several exciting angling encounters. He "hooked a 19 ¾-inch bass... weigh[ing] three pounds 10 ounces. Mr. Berringer had a cheap bamboo pole and a poor line, while the hook was of the cheapest variety. When he hauled out the fish he threw it back several feet from shore, when the hook broke, but the angler fell upon his prize and saved it. Berringer says he caught a pike more than three feet long a few days ago."[38]

It takes virtually no time, however, for the multiple-nature usage of a facility such as the Quemahoning Reservoir to take shape. By 1914, there were reports of sightseeing in the area and, most importantly, recreational fishing in the Que. While the winds of war blew stronger across the Atlantic in Europe, locals around the reservoir used it as a site to host picnics, family reunions and other gatherings. Newsworthy events, so redolent of the local cadence and location, were sometimes captured, as in the *Ligonier Echo* in 1915: "Mary Clawson was at Johnstown Thursday and Friday with her sister, Mrs. John Warren, and enjoyed an outing at Quemahoning Dam."[39] By 1916, the Quemahoning Dam was widely known enough to be included in a series of stops for a party and convention centered on Johnstown. Then-Mayor Franke welcomed the visitors, and they would spend time at the famous Inclined Plane, the steelmaking facilities and the areas around Johnstown and Windber, which included the Que. Another outing to the Que in 1916 included a number of attendees at a Pennsylvania "third-class city" league that enjoyed lectures on topics such as the civil service, municipal finance, electrification and a tour of the Quemahoning Dam as apparently the latest in engineering accomplishment. The *Altoona Tribune* reported other visitors and tours, such as in 1919, a feting of groups of soldiers back from the European theater with outings along the "largest lake in Pennsylvania."[40]

Nostalgia sometimes doesn't take long to form, either. In 1916, the *Ligonier Echo* eulogized the old topography around the Que in a tone that strikes one as jarring against the usual gee-whiz theme that would commonly be exuded in considering the wonders of the industrial age. The *Meyersdale Republic* noted similar sentiments:

> *Our journey led on to the Quemahoning dam. At the sight of that grand sheet of water our thoughts wandered back to seventy years ago, when our grandfather resided in a small village that stood on the ground that is now the site of the dam. A number of foundations of dwelling houses that were razed to make room for the water are submerged and one could only wonder how many people in the days gone by lived, loved, toiled and died on the land that is now covered by the Quemahoning dam.*

In addition to being the site of any number of family outings since the day the reservoir began to fill, it was a resource for fishing. By 1914, coinciding with the opening acts of the Great War, the Quemahoning, similar to Wilmore and Hinckston, was also a site for the stocking of fish. In those days, or at least in this instance, it took the advocacy of a local legislator (Warren Bailey) to persuade the federal fisheries to deposit fish in the reservoirs. The Quemahoning received—tadpoles. While they're perhaps not a game fish, there is an ecological component to the act of stocking tadpoles, either to act as prey for game fish, such as bass and pike, or provide other niche purposes in an ecosystem. At any rate, Wilmore Reservoir also benefited from an influx of these larval frogs, though Hinckston managed to secure trout.[41] Even if the purpose of the tadpoles alone was to yield a crop of frogs, one can imagine the chorus of frogs echoing out of these new lakes, in the shadow of the great mills and machines belching smoke, steam and fire.[42] One might also wonder if the lakes were the recipient of what one would call today "backyard ecology," where fish are removed from one body of water and placed in another, for future fishing purposes.

It is then evident that different stocking programs have been in place in Pennsylvania for many years, with Wilmore and Hinckston being frequently mentioned in public notices and in the prevailing experience of the time. It is also evident that the integration of environmental, or at least conservationist, approaches with that of the industrial infrastructure has been an organized undertaking for more than a century. Pennsylvania state programs have made adjustments and changed at different times; for example, by 1930, Wilmore, Hinckston and the Quemahoning all had different species added to them, along with other waters in the region. One report indicated that Wilmore would be stocked with catfish.

The clarity of the water at Wilmore Reservoir's outlet on display; this photo also shows the importance of riparian forest. *Author's collection.*

Of course, with the benefit of hindsight, we can see that these lakes are now areas of great fishing interest; it's important to mark the outset of their usage as natural resources when, at least in the case of the Quemahoning, it was less than a year old.[43] The parallel or dual usage of the reservoir was not necessarily a transition from the industrial supply to a recreational usage—instead, the user groups in the surrounds of the dam found alternate possibilities for this "shared" resource as soon as it was created. Reports of family fishing trips, in addition to being quaint portholes into the past in what constituted news, continued through the 1920s. From outings during the earliest days of motorized transport, to hiking and fishing—and with the state-enabled support of the latter—this technical system was also a "garden" amid the industrial smokestacks. An early article in the *Pittsburgh Post* in 1917 describes the "heavily forested" watershed of the Stonycreek and Quemahoning and the problems the combined watershed of the Allegheny River had been experiencing due to acid mine drainage; it's worth coming back to this and noting that this was in *1917*, during the First World War and the period of great industrialization of the United States.

Another instance of consideration of the dams in the flood-prone area such as the Cambria-Somerset region of Pennsylvania, weather consistently drew the attention of eyes watching the spillways. In 1916, a report out of Somerset County noted that torrential rainfall in early July raised the water in the reservoir by three and a half inches in the span of 90 minutes, with water eight inches deep coursing down the spillway. On slow days (or perhaps not so slow news days, depending on the paper and point), it was a common occurrence to report on the presence of a person, group of people or family fishing at the Que. One report from 1917 is striking, as the anglers had come from Turtle Creek, an 80-odd-mile trek that would not have been an easy drive during this time, right on the eve of the United States' entry into the Great War.[44] Still, the reservoir would be constantly maintained and adjusted, as MWC sought and received permission to install flashboards in the spillways at the dam in 1921, which would regulate the amount of water contained in the reservoir, essentially by penning greater amounts of water behind the dam, prior to being diverted over the spillways.

Flooding was not necessarily the only water-based concern for the area. In 1922, a drought descended on western Pennsylvania and created significant hardship for both city and rural dwellers alike. The *Wilkes-Barre Evening News* reported from late November of that year indicating severe water stress:

> *The most serious situation in the State appears to exist in Johnstown, where it was announced yesterday that unless the drought is relieved in three days approximately 50 percent of the Cambria Steel Company plant will be forced to shut down, throwing 8000 men out of work. Plans are being made to turn the water of the Quemahoning Dam, which supplies water to the steel works, into the city mains, as several parts of the city are entirely without water, and the large reservoirs that in normal times furnish a plentiful supply have been drained dry.…The water available in Johnstown now is only 9 percent of the normal supply. The Hinckston Run reservoir, which also supplies water to the Cambria plant, has been supplying water for domestic uses and is down to 24 percent. Several breaks in the mains have made the situation more acute.…Virtually all wells in the surrounding territory are dry. Several small mines have suspended operations on account of the drought.*

Indeed, even with the drought conditions in late 1922, the water continued to pour into the steel mills. The water flowed—that is, until the first of the reported breaks in the main waterline occurred, giving an early flash of emphasis to the perils of large technical systems and their maintenance. Bethlehem closed its Franklin facilities in Johnstown, idling some 4,000 workers, until the waterline from the Que could be repaired. Other mills, such as Gautier and the Lower Works, remained in operation for the several days that the line was repaired.

The usage of the industrial water supplies for drinking purposes forestalled significant hardship during that course of time. While this was the only instance of a systemic switchover that I could find, it still can be noted as being fairly remarkable for this, a tertiary use. Of course, 1922 was long in advance of the potability requirements under statutes such as the Safe Drinking Water Act, but the usage of it for population purposes was assuredly welcome and needed—and not without its own economic costs for the area, with the effect of idling the steelmaking and coal mining enterprises in the area. Still, there would be further droughts beyond the one in 1922 where the low levels of water in the Que would attract attention.[45] True to form in this area of the country, however, the rains would come again.

It is perhaps unremarkable that the lake itself became enough of an understood and well-known landmark in the decade or so beyond its construction that it was held as general knowledge and used as a landmark; its construction was, after all, extremely well-known to any reader of any area newspaper. More to the point, the primacy of the steel industry in the area virtually assured that each family member of each steelworker knew of its existence and heard about it, if not daily, then with some degree of regularity. It was, therefore, inevitable that it would also be the site of crimes—even gangland slayings in the times of Prohibition and the peak of organized crime, of which the Johnstown area had more than its share. Dice games, illegal booze, prostitution, cards, horse races, baseball bets—these are but a smattering of the ways in which vice was readily, easily and commonly monetized a century ago. Simpler, if tragic, human behavior also can be found in the history of the Quemahoning, as well as Hinckston and Wilmore; news reports of the body of Stephin Nagy, a suicide who also left behind an estranged wife and children, was discovered at the Que in 1927. In 1928, a building contractor named Leonard Miller was shot in the head and robbed; his car was "riddled with bullets." Whether accidents of fate or intentional menace, human history became more and more entwined with the lakes.[46]

One of the more colorful, if morbid, instances occurred during the peak of Prohibition, reported in the *Allentown Morning Call*:

> *A murder mystery confronted police today following the finding of the body of an unidentified man in the Quemahoning dam* [sic] *near here last night. Three bullet holes were found in the body and a rock weighing about 150 pounds was tied to the legs. The pockets in the man's clothing had been rifled and all marks of identification had been removed. Dr. HS Kimmel, county coroner, said the man probably had been dead ten days. He estimated his age as 32 and said he appeared an Italian.*[47]

And then—just a few months later, in the *Meyersdale Republic*:

> *The biggest illicit distillery…to be put out of commission in Somerset County since the enactment of the prohibition amendment to the constitution, was that raided one evening last week when County Detective Knepper, Officer WW Cupp and several assistants raided the home of Matt Moslanka, between Pilltown and the Quemahoning Dam. The officers found fifty gallons of moonshine liquor, ready for the market, 1300 gallons of mash, and four stills, the largest—a fifty-gallon plant—being in operation when the officers made their unexpected visit. The still…apparently had been kept going day and night. Moslanka had on hand a large quantity of rye and about 500 pounds of sugar.*

The cultural firmament of the Que is as varied as the history that carries us forward. Instances of sordid paths in sordid pasts are one element, something to titillate and thrill. On the other hand, the lake today has a renowned summer camp. The idea of a summer camp is not new, however—the area is also home to Camp Harmony, sponsored by the Church of the Brethren. In 1925, the *Pittsburgh Press* noted that "Camp Harmony is said to be unsurpassed as a vacation retreat." The Church of the Brethren would also lease the facility to other churches and faiths; it had obviously moved on from the rather sad recounting earlier regarding the last service held in a church that would be obliterated by the rising waters of the reservoir. A later, rather snooty report from the Meyersdale *Republic* expressed approval of the camp in a haughty tone: "Many young folks are given a pleasant outing there each summer, where they are kept under proper rules and regulations and given good moral and religious training."[48] In 1929, it was reported that the camp had been in operation since 1923 and that it was the largest such camp operated by the Church of the Brethren. In greater context and sweep of history, Somerset County has had a long history of attracting new churches and forms of worship, dating back prior to the American War of Independence.

Camp Harmony continues to conduct programs today, an interesting insight into historical and cultural continuity in the area, and adapted to the nature of the technical system by converting and adapting its own usage of this "industrial" water supply.[49] While the idea of steel mill ownership and resource usage remained an important and largely determinative factor over the next several decades, there was a continued and perhaps even increasing appreciation for the conservation and ecological aspects of the system. This is a repeated theme in the history of the system—an awareness of the importance of environmental sustainability is not something that occurred with the closure of the steel industry and the downscaling of coal mining. Instead, it has been present all along; a Pittsburgh *Daily Post* piece from July 31, 1927, described the efforts of a multitude of fishing and hunting groups in the Windber, Central City, Shanksville and Boswell area (all near or around the Quemahoning) for environmental and conservation purposes—planting nurseries, establishing hatcheries and managing the wildlife more generally.

By the early 1920s and in the aftermath of World War I, amid great national movements and internal migrations, Johnstown was poised to continue its shattering growth. Limiting this growth, at least in one sense, was the availability of labor, and even with that, significant tensions existed between labor and management in much of heavy industry, particularly steel. The Hinckston Run Dam and Reservoir became the setting for such an episode.

Draining the watershed in Middle and East Taylor Townships, just north of the city of Johnstown, is the 400-plus-acre and 1.1-billion-gallon reservoir of Hinckston Run. Hinckston can supply up to 5 million gallons of water per day currently, and its controlled outflow drops directly into the Greater Conemaugh River, as the dam is downstream from both the Little Conemaugh and the Stonycreek. Also notable for Hinckston on an easy circuit around the lake is a scenic waterfall that is listed as a must-visit by several hiking enthusiast websites. Pennsylvania Fish Commission reports that the lake is steep sided—which is normal for this area of Pennsylvania and this region of the country—and it is an ideal spot for bass and panfish.

IN SHADOWS OF SHAME

It is largely in the shadow of Hinckston Dam that one of the most virulent and shameful episodes in the history of the city took place. In 1923, Black and Mexican residents of the city were expelled following a shootout that eventually led to the deaths of four police officers. The settlement closest to the dam was a neighborhood called Rosedale, and it housed the coke ovens[50] for the Lower Works of the steel plant. Creating coke fuel is a labor-intensive process, with temperatures in excess of 3,000 degrees Fahrenheit, with high amounts of noxious gases burning, thick dust, debris and noise. Jobs around the ovens were among the worst in the steelmaking world, and the workforces in the areas tended to be on the lower end of the socioeconomic scale. Over the preceding years, Blacks and Mexicans moved into the area as a further example of rapid demographic changes that altered the texture and stances of both capital and organized labor.

The Rosedale neighborhood was the site of most of the inhabitants fitting that profile. Mayor Joseph Cauffiel ordered that all Black citizens in Johnstown who had been in the area for fewer than seven years leave the city; the order also applied to Mexicans. An example of a populist approach to those times and perhaps not foreign from the 2020s either, Cauffiel resorted to assistance from the Ku Klux Klan and his own secondary role as city magistrate to give further muscle to his rhetoric. Even by the standards of the time, Cauffiel's racist views were problematic, and this anti-immigrant, anti-minority approach backfired in a most spectacular way, as he lost the primary election that autumn, coming in fourth out of four candidates. Johnstown was the legal and constitutional laughingstock of the nation, with both the governor of Pennsylvania and the president of the United States taking note of the occurrences in Johnstown and evincing disapproval of Cauffiel and the actions.[51]

By around 1920, organized labor was a significant force in the economy and the public discourse. What had been vituperative anti-immigrant rhetoric 30 years before with the influx of eastern and southern European immigrants had now been retrained on the Black and Mexican populations relocating to the industrial North, fleeing the Jim Crow South. Steel mills such as Cambria hired groups of African American workers directly by agents traveling through southern cities and putting them on trains north to Pennsylvania.

The companies tended to deploy these new workers as strikebreakers; doing so inflamed the rank-and-file of union members, fanning the flames of ever-present latent racism with the economic incursion that it also represented. The reality in and around Johnstown is such that its latent xenophobia had been a sleeping monster all too easy to awaken over the course of time. In the 1870s and 1880s, when Johnstown first experienced

Cambria Steel coke oven workers, circa 1890, prior to the social changes that led to incidents at the Rosedale coke ovens two decades later. *Courtesy of the Hagley Museum and Library.*

its true trends of significant growth, rhetoric decrying the "otherness" of people from southern and eastern Europe was all too common in correspondence, everyday conversation and newspaper headlines. Insults that would later be reserved for those of African or Latin American descent were used on Italians, Hungarians, Poles and any others. Being able to focus hatred and rage on a group of people became immeasurably easier when there were no repercussions. Harried groups have no friends. They are all, in essence, utterly alone in a hostile new world. The otherness in that formula was complete, and there was no opprobrium meted out to the bigots. The complete innocence of these groups was utterly beside the point when one needed to sow desperation, anger and righteousness.

Dave Hurst referred to the area around the Hinckston Reservoir as possessing some of Johnstown's most "profound" history, starting with Hinckston and the violent altercation between its namesake and Joe Wipey, the Indian he murdered, but also being the site nearest to the unethical, unconstitutional, illegal and grotesque events around the neighborhood of Rosedale and its coke ovens, with this forced eviction of thousands of Black residents. Hurst ably described the look of Rosedale: gritty, dirty and harrowing, with a couple hundred coke ovens producing coke and other important by-products targeted toward steelmaking, with thousands of residents, as he put it, "crammed into substandard Bethlehem Steel company housing." The contrast between what was there and what happened there compared to the hollow beneath the dam now is also harrowing and striking: "Here in a hollow named for a murderer, life is slowly reclaiming its birthright."

While the chapter somewhat closed on this when Cauffiel left office, the hundreds or even thousands of people who were forced to leave undoubtedly fled with fear and trepidation. The despicable acts left scars on the people in the area that would take decades to heal—if, indeed, they ever have. The theme in this work, however, is to keep in mind the strength and power of Big Steel in cahoots with a useful idiot holding political office and exhibiting just a hint of cupidity and thereby opening the door to first-rate corruption and abuses of power. It would happen again and again, especially a decade and a half after the events in Rosedale.[52]

As mentioned throughout this work, the question of maintenance surrounds the discussion involving the dams and reservoirs. Routine inspections of the dams had been in place virtually since their construction, certainly by internal teams at MWC and Cambria Steel but also state enforcement agencies as well. "'All of these dams are sources of potential

Rosedale coke facilities, circa 1945, aerial view. *Courtesy of the Hagley Museum and Library.*

danger unless properly maintained, and some of them store so much water and are so located above centers of population as to cause untold loss of life and tremendous property damage,' state[d] Charles E. Ruder, Chief Engineer of the Water and Power Resources Board. This Board also has supervision of the construction of all new dams within the state under permits required by law," as reported in the *Meyersdale Republic* in 1932. Information in 1934 from the *Duncannon Record* indicates an increasing level of scrutiny on such dams statewide, noting that the jurisdiction for inspection of dams was within the Department of Forests and Waters, with 300 dams in the state on schedule for inspection in 1934. The Que was one of several dams mentioned given its position vis-à-vis Johnstown and likely its strategic importance to a company deeply intertwined with the economy of the nation amid rapid economic and social change.

Change is, after all, a constant—and it is both a dynamic of power and a reality of the existence of a large technical system. Some changes are social; some are environmental. Others reify the constant, if not constantly understood or observed, undercurrent of continuity in the passage of time.

And yes—some others offer catastrophe in the making.

A Catastrophe That Was and Was Not

As the genial reportage of family outings to the reservoirs gradually faded from coverage in the local news, a new terror fell in the area on March 17, 1936: a great flood.

An early international report came in from the *Vancouver Sun*: "A wireless dispatch from the United States naval station at Johnstown at 2:39 this afternoon stated: 'Dam breaking. Town will be wiped out. We are moving to higher ground.'"

Thus lays clear and bare the recurrence of all the fears stated from thirty years earlier, that the Que (or any other dam) was going to be the cause in fact of Johnstown finally being wiped from the map. What the much smaller South Fork dam couldn't do in 1889, the state's second-largest lake would accomplish.

The weather pattern associated with the St. Patrick's Day flood of 1936 was a combination of a sudden warm spell that attacked a deep snowpack that had formed over the weeks and months prior as the region began to emerge from a notoriously difficult winter, even by local standards. The combination of rain and the sudden thaw created conditions all over western Pennsylvania that led to widespread flooding. Nor was the flooding limited to western Pennsylvania; the weather pattern that had started this event in February led to flood conditions up and down the Potomac River, with towns between the Appalachian Mountains downstream to Washington, D.C. all receiving heavy damage. Pittsburgh's river waters crested over 45 feet, swamping the downtown buildings. It is still, to this day, the single largest flood event in the city of Pittsburgh. It was not, therefore, only Johnstown that was affected; but it was only Johnstown that had the legacy *and* the dams. From the *Ponca City News* on the day:

> *Major Lynn Adams of the state police told his office today he was informed that the Quemahoning dam in Johnstown "is leaking badly and likely to got* [sic] *out at any minute." Adams said the report came from a man within sight of the dam, which holds nine billion gallons of water. Later Adams said the information concerning the dam came from one of his troopers within sight of the dam. Major Coleman B Mark, superintendent of the Mt. Gretna military reservation, reported to Harrisburg later that an operator of an emergency radio station, set up by the national guard, advised him at 3:20 pm that water "had burst through" the Quemahoning dam, but that the main structure still was intact. The station was within sight of the dam, Mark said.*

It is difficult to fully picture what was being described. The articles are based on dispatches from many hundreds of miles away, a fascinating picture into a newspaperman's game of telephone. The Quemahoning Dam never breached during the events in March 1936. If, for example, the dam was "leaking badly" or water had "burst through," it is likely to mean that such an earthfill dam would be doing what was called "piping," a potentially and nearly unavoidable catastrophic loss of integrity where the water penned upstream by the dam was pushing through weaknesses in the dam itself, exerting many thousands of pounds of energy on fill that is now loosening and no longer capable of holding in place, such as at the South Fork dam in May 1889.

The 1936 flood in Johnstown was not occasioned by any great dam failure or a catastrophe of biblical proportions. It was, in fact, the same type of flooding that had occurred everywhere in the region at that point. Some 20 people were killed in the flood itself, and many dozens of buildings and homes were damaged or destroyed; total damages were estimated to be around $41 million, or about $1 billion today. The widespread destruction of the regional flooding was amply recorded; it included hundreds of thousands of families cast out of homes, the Golden Triangle of downtown Pittsburgh submerged, all transportation and electrical power generation halted. It was, therefore, a significant event, but Johnstown was not singular or special as a result of having been victimized by yet another flood.[53]

What then to take away from this as a lesson, when viewing it through our lens of a large technical system, the importance of maintenance and, at least in this small subsection of time in March 1936, both the importance of communication and the reckoning that breathless and inaccurate reportage can pop up at any time. This might be laid at the feet of newer technology (which is discussed more fully in reviewing the Wilmore Dam situation to follow), but that too is nothing new. Indeed, the editorial of a Canonsburg, Pennsylvania newspaper closer to the events as they actually occurred does a bit to root for the home team:

> *This newspaper has paid tribute to the important part the radio played in the recent flood catastrophe. But it feels called upon to point out one weak spot, which has to do with the operation of radio broadcasting news in general and flood news in particular. The newspapers were not entirely free from the false rumors and inaccurate flood news. But the radio was the principle* [sic] *offender.*

The competition between legacy and emerging types of news (or information) delivery is something that we live with every day in current times, with the omnipresence not just of social media but of the internet itself. The 24-hour news cycle is something that we, in the 21st century, have lived with for decades; it's likely that the instances of the competition between the newspapers and the radios (and then subsequently, the television, which broadcasts moving images of any type of horrific nature that people will imbibe with rapt attention) in the 1930s led directly to the phenomenon that we see today: half-baked stories which have a salient political vector that rockets around the world or through a particular community, forever freezing the stance into a polarity. It is hard to ignore the parallels of today as we see the Canonsburg scolding continue:

> *The newspapers did nothing, for instance, as serious as the action of the National Broadcasting Company in picking up the statement of an amateur broadcaster from Johnstown and giving to the world the news that the big Quemahoning Dam had burst. Not stopping there, over-zealous broadcasters immediately sent word over the nation of the terrific loss of life and property as a result. The newspapers, whatever their faults may be, have developed a technique of verification which would make that sort of thing impossible. "I heard it over the radio" is the all too frequent description of information that can't be accepted until it is verified by the newspapers.... Individual newspapers and the press associations have built up a system of investigation and authentication as well as of news gathering which is designed to get the news of the earliest instant and to check it at the same time. There is always a method of backchecking to verify the news and before it is given to the public every effort is made to prove the facts. The radio...is too prone to accept rumor as facts.*

No matter what direction or stance one might have on the fealty of any form of media to facts as they eventually emerge, it is impossible to not see the emerging rivalry here in the breathless reports and then the answering capstone story a couple of weeks later, given the opportunity for more information to unfold and for that type of verification. Fast-forward a few decades, as we'll see, there will be a new instance of speeded communications hyperbole. There is clearly a need for urgent information when such a situation would warrant; the difference is in being able to discern when that need is and, in fact, to correct a story as soon as possible. Still, while the Canonsburg editors' scolding might be taken as a fist-shaking trope, the panic and error in reportage were quite real. Another newspaper, in Stockton, California, noted that the report of the failed dam near Johnstown had come from an amateur radio operator, who "narrowly escaped" drowning and was rescued.

It may well be important to take a bit of a subtler note involving communications than what the Canonsburg editorial writers did. Even in 1936, for example, there was not just "a" radio station. There were multiple radio stations and channels, and some radio operators, particularly those affiliated with the War Department, were acknowledged and saluted for their contributions to facilitation of full information. A laudatory example hails from the *Gettysburg Times* in April 1936:

> *It has been brought to my attention that, as radio operators of the Army-CCC station WVH and amateur station W8MRI, of the army amateur system, you were instrumental in keeping the War Department informed as to the correct situation regarding the conditions of the Quemahoning dam at Johnstown. When all communication with the outside world was cut off as a result of the recent serious flood conditions you, through your devotion to duty, kept your station continuously open during a period of 20 hours and were the sole means of communication with the outside world. I am pleased to commend you on the excellent manner in which you furnished this information.*

St. Patrick's Day flood, downtown Johnstown, 1936. *Courtesy JAHA.*

The story of the Quemahoning, along with Hinckston, Wilmore and the smaller dams in the service of MWC during the 1936 flood is exactly the nonevent of the dams holding and continuing to function according to their design. This part of the story, then, is not about catastrophic loss and damage but instead is a nod toward design, construction, maintenance and some degree of luck. The everyday story that could be run—by newspaper, by radio or by social media—is one that no one really cares about and yet they should: the dams held, and tens of thousands of lives were saved.[54]

No stories of catastrophes averted should be expected, perhaps, but there was a celebration of a sorts, announced in early April 1936, that a statewide inspection of dams that withstood the St. Patrick's Day flooding showed that the biggest—such as the Que—were in "splendid" condition. The survey was completed by a group of state agencies, including the Water and Power Board and the Department of Forests and Waters, and also filled in some blanks on the mid-action and after-action report of the Quemahoning Dam performance during the peak of the flooding. The data revealed that the "water never rose to a height of more than 5½ feet in the 12-foot spillway, and the amount flowing was a little more than one-fourth of the capacity of the spillway." If this is indeed the case—and it seems to have been—one encounters one of the truths about panicked reports from supposed eyewitnesses; they are often wrong. How can that be? How can there be panicked reports about an enormous dam possibly collapsing, in its death throes by piping, and thereby triggering each and every vivid memory of what had happened at South Fork nearly fifty years earlier, but with the dam *actually functioning normally* during the unprecedented onslaught—and not only functioning normally, but actually quite distant from maximum capacity?[55] It may be that the appearance of even a few feet of water plunging down a spillway might give the appearance of impending doom. It would then be *that* report that was the juiciest and therefore the story that captivated listeners and advertising dollars.

The remarkable aspect is not necessarily the erroneous reporting or the panic but the uneasy, almost liminal, existence that the system has in the awareness of most of those of us who are not directly charged with ensuring its continued and smooth operation.[56] It was not all without any impact at all, however; a subsequent visit to the area by President Franklin Roosevelt included an auto tour of the Stonycreek River and the Que, all of which were a part of the persuasion package that triggered the channelization project in the area. A month after FDR's visit, the state water and power regulators announced that MWC had applied for a permit for repairs on the Quemahoning Dam and the construction of retaining walls in the Johnstown area and in 12 other Pennsylvania counties.

When the Pittsburgh press run headlines such as "BIG DAMS HOLD UP" and proceed to extoll the strength of the dams around the state, most especially the Quemahoning, either because of its size or proximity to Johnstown with that city's flood history, one gets a sense that technology is turning the corner against nature. This is, of course, an open question to which we shall return; turning the corner is not likely, not in the grand scheme of time. As we noted at the outset of this story, Pennsylvania was once center-continent and equatorial in climate. Nature moves as it does, with or without human influence. However, it is possible to interact with nature in such a way that preserves life, limb and property. Without maintenance—and its attention to detail, its unsexy "plodding" attributes, its rather thankless and never-noticed work—nature can turn a corner far more quickly than technology can. This also seemed to be present in the minds of many. While the 1936 flood was notable—because it happened—the following year, enough rain had fallen to again warn of possible floods.[57] Indeed, toward the end of World War II emerged news reports of great rains causing streams and rivers to swell, including the Stonycreek River, which swamped basements and resulted in the drowning of two children in Hollsopple and Davidsville. The news noted the flooding conditions in southern Somerset County as well, with 10 feet of water swamping Route 40—the National Road—and flood-control measures taken near Confluence, at the Youghiogheny River. Gravity, hillsides and water were never far from the awareness of the people living in the shadows of the dams.[58]

3

STEEL POWER AT ITS PEAK

1937–1977

As events moved toward the 1940s and the Second World War, however, we can note that the placement of the dams and reservoirs in the area again became a focus of attention, albeit from a different standpoint and in a different direction. At once a target for worry and trepidation, along with being the irreplaceably important—if invisible—connection to a natural resource not typically factoring into the steelmaking process, and an ongoing and growing site for popular recreation and presidential sightseeing tours, the importance of the water supply would continue to take different and important shapes for different public uses, legal and extralegal alike.

Not a "Little" Strike

In 1937, the Que furthered its stance as a cultural reference point and as a soft target. In one case, it became a landmark to visit by a group of hikers—the Provincial Hikers—who seemed to have routine programs of walking and visiting sites of cultural and historical significance. Different members of the Hikers put together educational lectures or programming, including photography and slides; in one instance, the Quemahoning Dam was the location of interest for their route of cultural and historical education, along with Fort Dewart along the old Forbes Road at the border of Somerset and Bedford Counties. While the Que was taking its space in the foreground as an area of historical significance even though it was less than a quarter century old, it was also in the background but then thrust violently forward as a threatened resource in struggles, as opposed to celebration. The mills and the lakes again became national news, as reported in Wisconsin in June 1937:

> *Mysterious blasts broke two pipelines supplying water to Bethlehem Steel's Cambria works early today. General Manager CR Elliott said it might be necessary to close the works for as long as two weeks. The mills had been reopened a week ago after being closed by martial law....To prevent damage from a sudden stoppage of water, Ellicott ordered five blast furnaces banked and closed 15 open hearth furnaces and the rolling mill in the Franklin plant. The first explosion about midnight broke the six-foot line running from the ten billion gallon Quemahoning Dam, 11 miles south of Johnstown, to the Franklin mill which consumes half of the water used by the Cambria works....The second blast at 3:05 am...was heard for more than two miles. It snapped the 36-inch line from Border dam to the Gautier mill, second largest division of the works.*

As might be expected, the explosions caused a rapid loss of water and water pressure, depriving the steel mills of the ability to operate. Steelmaking, as always, used millions of gallons of water every day of operation, including as an important coolant for furnaces and other machines that were handling and manipulating molten and very hot steel, with temperatures exceeding 2000 degrees Fahrenheit. The two plants named in the article, Franklin and Gautier, were half of Bethlehem's Cambria footprint, with the Franklin facilities being exceptionally large and in operation since around 1898.[59] In short, if these plants were not in operation, Johnstown wasn't in operation, and if Johnstown wasn't in operation, Bethlehem was suffering.[60]

Whoever planted the bombs, then, knew exactly where to hit the company.

What was the reason for the sabotage? The answer: The Little Steel Strike—except that there was nothing little about it.

Earlier in 1937, the country's largest steel company, USX, had executed an agreement with organized labor to formalize new working conditions, including an eight-hour workday, a uniform pay scale and other benefits, such as increased pay rates for overtime work. While USX went forward with the agreement, the smaller steel companies did not—hence, "Little Steel" and the Little Steel Strike. Bethlehem Steel, despite its size (and absolutely the largest steel company in the Johnstown area), was far smaller than the gargantuan U.S. Steel. Other steel companies, based in Youngstown, Chicago and elsewhere, all had significant strife. In fact, it is rather unfortunate that the name of this national labor action has been labeled "little," as there was nothing small about it. It was the last of the large-scale, violent strikes, with thousands arrested, hundreds injured and dozens killed.

The Little Steel Strike could be one of the most underrecognized episodes in the history of the U.S. labor movement. Certainly, documentation and even films have been made about any number of other labor strikes in the late 1800s and first half of the 20th century, most especially about coal miners and the formation of the United Mine Workers. For unknown reasons—and I cannot dismiss marketing, as the name alone of "Little Steel Strike" might not resonate as something important, even to serious students of history—this seems to have been the case. It is, therefore, useful to recapitulate the major events nationally with the strike and also explore some of the local effects on Bethlehem Steel's Cambria facilities, as they were a major location for the action. It also underscores the pressure points of power of this techno natural system and its spot in regional and even national transformation.

The Little Strike is best known for what was labeled—then and now—a massacre that occurred near Republic Steel facilities outside Chicago, Illinois. A group of around 150 police patrolled the grounds around the facilities, facing off against picketing workers. As events turned out, the police fired on the strikers; reports noted that several dozen people were shot, 10 of whom were killed. Apparently unsatisfied with those results, the police proceeded to beat around 100 more with clubs. Thus, one of the most notable events in 20th-century labor relations was the Memorial Day Massacre of 1937.

Bethlehem's Johnstown/Cambria facilities were a major prize for either side. For the company, it was an undisputed massive plant that provided a significant portion of the daily production of the entire organization. For the workers, it was largely the only game in town in terms of employment, unless one wanted to move to a coal mine (with a good chance that there was Bethlehem ownership as well) or the much smaller U.S. Steel plant located just up the Stonycreek River and therefore was already a part of the new labor agreement. However, in the valley isolating Johnstown from much of the rest of Pennsylvania, Bethlehem was the elephant in the room, with more than 13,000 workers in a city of roughly 70,000.

Despite the direct employment footprint of nearly 20 percent of the entire population (and this is without factoring in other entities, such as public works, the smaller USX operation, other non-Bethlehem coal mines), the practices engaged in by management were often far from reasonable, let alone equitable or fair. Some workers would show up only to be sent home a majority of the time—resulting in no pay. Others were required to perform maintenance and upkeep at a foreman's house—or lose their jobs. Other times, a foreman or manager would request meals to be prepared for them and brought to them at the plant—for no pay. One famous example included a request to marry the worker's daughter, and when the foreman's entreaties were refused, the worker lost his job.[61]

As the spring of 1937 moved into summer, those working for one of Bethlehem's captive railroad companies, the Conemaugh & Blacklick, moved to strike. This resulted in a general walkout throughout the entirety of the Cambria facilities, stilling the operations and fully engaging their participation in the Little Steel Strike.

These events, naturally, did not allow for the amicable resolution of differences.

While it didn't rise to the level of the Memorial Day Massacre in Chicago, the Johnstown strikes did result in rioting and violence. Police fired warning shots and tear gas to disperse the striking workers picketing the mills. Soon after, given the connection between the burgeoning steel union and that of the United Mine Workers, coal miners also walked off the jobs.

Johnstown has a long history of colorful mayoral characters. As noted earlier, "Fighting" Joe Cauffiel was a morally crippled fool and attempted—and partially succeeded in—the forced removal of the Black population of Johnstown in 1923.[62] In 1937 however, Johnstown still suffered under a similarly half-baked dolt and blowhard named Daniel Shields, a man grown to wealth and local prominence through trading in illegal alcohol during Prohibition and an apparent lackey of whomever opened the largest billfold. In Johnstown, in 1937, there was no question as to who (or what) held that billfold.

Mayor Shields—the best mayor money could buy—aligned his interests firmly with currency rather than constituents and brought the "strong mayor" model of city governance to bear on the protesting workers, joining the Pinkerton thugs that Bethlehem already employed to harass and threaten the protestors. Shields "deputized" hundreds of citizens—untrained as police or law enforcement, even by standards of the day—to join the security around the plants and attack and disperse protesting workers, which Ahmed White, in his history of the strike, rightly called "gangs of vigilantes" on missions of intimidation and threats around the city. In addition, Shields, like Cauffiel in 1923, also took advantage of another foible of the mayor's legal authority of the time, which was to sit as a magistrate in a municipal court.[63] In the execution of those important duties of office, he cited picketers for littering, fining them exorbitant amounts ($100, which would be over $2,000 in 2023) and even imprisoning them for months. Mayor Shields, having already issued arms and nightsticks to his roving gangs, told the "union people that he planned to prevent Bethlehem from importing strikebreakers, provoking the strikers, or even running the mill during the strike, as long as the picketers remained peaceful. He lied."[64]

The situation escalated, with both physical violence and the mayor issuing astonishing propaganda pieces to local media, including a daily newspaper and radio stations. The governor had already dispatched hundreds of state police, and the mayor felt that even more muscle and fabricated claims and misinformation were needed. Much of the resourcing for channeling the mayor's rage on the public stemmed from Bethlehem itself, contributing more than $25,000 (likely more, according to Ahmed White) to the cause over the span of only about a month. That is, in today's terms, more than $530,000 allotted solely for the forces of the government to attack, arrest, detain, intimidate, brutalize, imprison and terrorize the workers and the population of Johnstown.[65] Bethlehem Steel then was always ready to open its pocketbook when it came to dealing with organized labor. With this as the backdrop, the only true surprise about the fact of a desperate group of strikers attempting to destroy the water pipelines to shut down the mill is that it was apparently limited to only two attempts.

While the reports of the rioting and violence in Johnstown were somewhat exaggerated, they were more than enough to draw the further attention of the governor in Harrisburg, along with politicians in Washington, D.C. The governor, short on patience with the preternaturally incompetent Mayor Shields, sent hundreds more State Police to Johnstown and ordered the mills closed and evacuated until the labor situation was resolved. The governor actually threatened Shields with arrest during all of this, and Bethlehem mounted enough return threats to maintain around 1,000 workers to continue to be stationed at its facilities. Through a series of non-trials and significant errors, when the governor eventually lifted his order stilling both sides, there was a limitation on the number of picketers allowed to be present, which worked to de-power the labor forces while perhaps unintentionally strengthening Bethlehem's hand. As reported by the *Mount Carmel Item*, the deteriorating situation continued:

> *The dynamite blast which tore ragged sections from the 62-inch main…*
> *helped close down the strike-harassed Bethlehem Steel plant at Johnstown,*

> *Pa. Another explosion damaged another and smaller main. The two mains daily carried almost a million gallons of water into the plant. The dynamiting came after the plant had opened under the protection of martial law. Strike leaders "deplored" the sabotage as strongly as public and plant officials.*

Bethlehem acted quickly to repair the damaged pipelines, sending out trucks of materials even as the gouts of water spread across the area. This apparently caused even greater friction within the labor forces, as the strikers were continuing to take the labor action while the company itself worked to repair the water lines. The damage to the lines included a 20-foot-long rip, spilling thousands of gallons of water out into the tunnel through a hill. While the damage occurred, was assessed and brought to repair, the Pennsylvania State Police sent 200 troopers in to assist in the peacemaking efforts around Bethlehem's Cambria facilities.[66] The troopers were charged with stopping and searching all vehicles coming in and out of the area at the time, presumably looking for firearms or materials that could be made into sabotage devices. Investigations determined that dynamite had been planted between the metal pipeline and the concrete work of the tunnels surrounding it, as both the large gap in the pipeline existed and the concrete supporting it was cracked and damaged. As befit a company of its resources, influence and importance, Bethlehem's response was decisive and thorough, and the facilities were not out of commission for long, despite the significant damage.

Taking a step back for a moment and considering these events in the context of the large technical system and the networks of power, we encounter areas that catch our attention and urge us to reflect. Clearly, this incident demonstrates the case study of the system-within-a-system situation of the water, dams, pipelines and reservoirs and their relationship to the larger steelmaking network. All these physical aspects are embedded not just in the earth but within the framework of the knowledge that it takes to run a complex industrial system. The infrastructural element is a key component; invisible and buried in tons of concrete and tucked into mountainsides, it still represents an essential input into the industrial complex whose presence is immediately noted when struck by a saboteur.

At the same time, this infrastructural input is the pressure point in what would become an increasingly violent and perilous mass labor action. In this context, we are not just referring to actions as legal or not; we are referring to actions as they were taken, sometimes and perhaps without full regard to what would have been within the ambit of "legality." The presence of the water here is a euphemistic reservoir of energy, of pressure, of leverage that acted as a buffer and as a threat. It is the context of the action alone that would change the nature of the water's power in such an instance. While the interests of labor did not carry the day in terms of the outcome of the Little Steel Strike—beyond the harebrained and corrupt machinations of Mayor Shields supported by a state governor on alert against insurrection and bankrolled by an enormously powerful and wealthy company—the strike collapsed. Eventually, with the advent of the Second World War and the industrial policies adopted to develop and supply the nation's warfighting capabilities, workers did succeed in attaining many of the parameters that we all take advantage of today: eight-hour workdays, elevated rates for overtime, paid time off.

As things moved forward, Bethlehem Steel was found to have been engaging in unfair labor practices and ordered to stop exerting force through miniature tyrants like Daniel Shields. The mayor himself suffered little as a result of his bloviating and, frankly, illegal strong-arm thuggish tactics. He was called to testify before Congress and admitted to destroying records (including employment records of the roving gangs he had patrolling the streets of Johnstown). He then fades into history with little consequence, other than playing the role of dimwitted and loudmouthed henchman who won the battle but, eventually, lost the war. Congress mulled all of this over, including the pipeline sabotage, in the hearings investigating the Little Steel Strike and its aftermath.[67] There would be no personal justice, except for that of the arc of time, when the forces came together to defeat the darkest days of Europe and the far Pacific.

Time marches on. As the Little Steel Strike fizzled at the end of the summer of 1937 and as ominous clouds rolled across the skies of Europe, life returned to something akin to normal in this industrialized garden in the hills of western Pennsylvania. Panegyrics about observing the beauty of nature around the Quemahoning to letters using the Que as an example of the ever-present fear of disastrous flooding—the same battlelines drawn decades earlier remained steady in an uneasy rhetorical truce straddling the sublime and the catastrophic. Tales of great waterfowl hunting expeditions in November 1940 along the Que, with some estimates of swarms of birds in excess of 15,000 geese and ducks on the Que alone. These stores and this reality vied for a growing awareness of the dams' potential as a wartime target and prize. On the eve of the United States joining the Second World War, policymakers in Harrisburg again ordered the inspection of all the

Disgraced Mayor Daniel Shields testifying before Congress in 1938, in the aftermath of his perfidy with the Little Steel Strike in 1937. *Courtesy LOC.*

dams in the Commonwealth. Around the same time, greater precaution grew with an awareness of the threat of war and sabotage; the MWC had requested greater police presence over the Quemahoning, with a local judge appointing three individuals to do so. In addition, several individuals were hired to patrol and watch over watersheds near the reservoirs and dams.

As to the way things looked at this point, with huge steel mills pouring smoke and blasting iron into steel, it must have seemed that heavy industry had an eternal momentum, for the owners and for union members. Most of the history of big labor in Big Steel through the post–World War II era is one exemplified by the pictures that one would see of those postwar times. Steelworkers and their families had great affluence, far more than what the previous generations might have dreamed. Mammoth cars shining with chrome and made with steel many of these workers had helped make plied their way about town on streets crowded with pedestrians, shops and streetcars, performing their patrol after hailing from well-kept and spacious houses. Children of this era went to college in droves, particularly in the

1960s, as the baby boomer would become the flower child. Benefits that accrued to Bethlehem Steel workers in the 1950s included six or more weeks of paid vacation, shift availability preferences and wage differential with seniority, fully vested pension rights and medical care. Times were prosperous, and times were good, during this sometimes-uneasy trilateral alliance between government, industry and labor.

A couple of years later, the Que was "targeted" in a practice air raid drill, with a successful sneak attack consisting of planes diving from a great height and destroying the dam, unleashing the billions of gallons of water, flooding Johnstown and the vicinity and, perhaps worst of all for the warfighting effort, knocking the Cambria steel facilities out of production. Bethlehem Steel, as an entity, carried enormous importance for the United States during the Second World War, supplying armor plate and naval guns. Most importantly, Bethlehem owned more than a dozen shipyards, which cranked out merchant vessel after merchant vessel, providing the instruments for the supply of a truly global war.

Revisiting the Environment

Despite the obvious productive capacity, and Bethlehem's work in contributing to the single-minded national determination to see things through to ultimate victory to conclude the Second World War, it is worthy of note that other facilities Bethlehem built presaged the eventual direction of the Que and other reservoirs to their use today. Amid all the environmental challenges known to be caused by the effects of coal mining, acid mine drainage stands out and is still much of the legacy in Pennsylvania, West Virginia and elsewhere today. But in 1944, in, of all places, the major daily in Pittsburgh noted the following:

> *Thus a far-sighted industry saw the handwriting on the wall many years ago; at no urging from outside interests it went to the expense of conducting experiments which proved definitely that acid mine drainage could not only be neutralized and rendered fit for industrial and domestic consumption, but it could also be accomplished at a profit. If these polluter-sympathizers wish to be convinced of the emptiness of their claim that there is no known practical method of treating acid mine drainage, then let them visit the neutralization plant of the Bethlehem Steel Co. at the Quemahoning Dam,*

in Somerset County, where the acid-laden water must be treated before being used in the mills at Johnstown.

It seems striking that this piece was published not even three weeks after the D-Day landings at Normandy, France. It is at least notable as well that the construction of the writing and the argument also pit a major industrial force—Bethlehem Steel—as a challenge to other industrial operators. It is, however, unlikely that Bethlehem Steel took the same precautions around all the many coal mines that they too owned in the area—those same mines today still leach waste into the watershed of the very system examined in this work. Back in 1944, that was unmentioned; of course, it is also arguable that the writer could claim that the imprecation goes directly toward Bethlehem's own stewardship. After all, if it can be done there, the argument would say, it can be done anywhere. The costs of acid mine remediation are quite high, no matter what era. As we've seen, acid mine drainage had been an environmental concern since at least the middle of the *First* World War, decades prior to the 1944 report in Pittsburgh. Clearly, however, the writer of that article could see the great costs that would be the legacy of heavy industry and the burdens to be borne by future generations. At the same time, however, those future generations were to be afforded the prosperity made possible by the industry; again, a push and pull, an uneasy truce, a wobbling and spinning top.

Still, this mention amid the ongoing dark clouds of World War II is not the first public expression of concern regarding environmental contamination, nor were coal mines the sole culprit. More than 30 years earlier, in 1916, there was a public demonstration of indignation and worry from the angling community pertaining to the pollution in the Raystown Branch of the Juniata, near Bedford, caused by the outflows of chemicals from area tanneries. It was not, therefore, just all about acid mine drainage. The piece focuses on waters not a part of the system that is the subject of this work, but the salient aspect here is the community of users of a natural resource was already well-established enough to carry reportage on environmental degradation in the area before the entry of the United States into the First World War. The usage pathway—angling, outdoor recreation—is the one constant area of continuity that arises time and again, becoming deeply entwined with the Johnstown system itself and the greater region surrounding it. The environmental concern was, therefore, always present, just as stocking fish was always present. The concern around the time of the Second World War—call it early awareness conservation—used different channels, and of course, the volume and tenor were much different than what we might see today. The underlying dual threads of outdoor usage by different groups of the public and the more existential concern about environmental degradation were coexistent with the system. The changing of times and circumstances surrounding the system created different levels of awareness and different areas of emphasis, but the awareness was always present.

Hinckston Run, like the other bodies of water in the system, continued to be a site for area anglers to enjoy since its construction and the stocking of fish. Reports of fishing and hunting opportunities around Hinckston date to as early as its construction and continued throughout the decades of Peak Steel in the area. In 1945, some 30 years after the reportage regarding the establishment of fishing derbies and shooting contests in the area, the *Shamokin News* carried a brief story of high school freshman Ron Hill, who "landed a 21-inch black bass, weight three pounds, in Hinckston's Run Dam [*sic*] near Vinco."[68] This story, barely three weeks after V-J Day, continues to lend credence to the view that while Bethlehem marked the property as private, the open and frank usage—and news reporting—of such stories continued to show the industrial system as a natural resource.

Hinckston Run Reservoir from the north end, looking to the southwest, with excellent water clarity and riparian forest. *Author's collection.*

By the late 1940s, the Que had been in place long enough to see a generation born alongside of it grow into middle age (and, by the standards of the time for metrics such as life expectancy, perhaps older age as well). As such, it had become a natural example of a legend given to longer-term weather prediction. One colorful wag, in the *Somerset Daily* American, noted that the use of the word *dam* was "noun, not an adjective" when discussing the possibility—or lack thereof—of a big snow coming to the area, because the commonly held belief was that the winter would forgo heavy snowfall until the reservoir was full and, as of December 1948, it wasn't. Thus underlies another thread of continuity that is largely present throughout all of society and not just the Johnstown area—the connection of an awareness of the natural world to the needs of agriculture. This link is, indeed, timeless but reflects another dimension of the primacy of the system. Its very nature becoming a part of the environment and, in a sense, disappearing is actually evidence of its importance. It is so large, and so much a part of daily life, that it isn't seen as separate from either the industrial or natural world.

Indeed, the entwined combination of history and nature came up during a statewide autumn celebration called Pennsylvania Week, reported in the *Daily American* in October 1949:

> *The first historic and scenic tour will be made today to the Quemahoning area.... The trip will be made in a bus provided by the Somerset Bus Company at operation cost plus taxes as that company's contribution to the Pennsylvania Week program.... The tour will go past Stonebridge, the farm of John Zimmerman to the west side of the Quemahoning dam which will be seen from the east side after the turn in made in Hooversville. The site of old Fort Stonycreek, near Kantner, will be pointed out as will the former place where an Indian, Kickenapaulin, of pre-revolutionary times held sway. The old Forbes road will be crossed and re-crossed on this tour.*

This may well have been the first historical "tour," but it was not the first signature of a new interest in the history of the area. Paul Trimpey, writing for the *Somerset Daily American* in 1945, commemorated the entire county with a high-level history of its 150 years. Starting from pre-colonial and then colonial times, Trimpey noted the abundance of natural resources, the agricultural opportunities and the county's venue for "escape" from other parts of the region. By the time of his writing, of course, the massive steel plants and coal mines strewn all over southwestern Pennsylvania were at their absolute height in terms of manpower and production, and Somerset

County, while representing more than its own share of coal mining, was somehow different and apart from the rest of the industrial scene at the time.

We see from these examples important glimpses of a renewed interest in the past and the places in which events were situated. The dam and reservoir might have been well be too new or well-understood to be anything other than a waypoint on that journey. Of course, we don't know what was actually discussed on that bus tour in 1949, but even if the history of the dam itself was ignored, with it being less than 40 years old then, it is still an important point to notice its background as a critical piece of infrastructure in the area. Certainly being seen as a stop on what was intended to be a "scenic" journey, much as it had been for years prior, also showcases its role in an increasing awareness of the natural topography and beauty of the area. Parallel to this is that this shows not just a letter written to a major city daily in a persuasive tone warning of environmental damage but also that the reservoir is being seen increasingly as a *lake*.[69] This is more than just a difference in terminology; it connotes a different usage and understanding of what the thing actually is. Indeed, the trend would continue, as reported in the *Somerset Daily American* in 1971, when the Forbes Road Association "toured a section of the old trail between [a home in Stoystown] and the Quemahoning Dam....Under the expert guidance of [historian Fred] Doyle [the hikers] were able to ascertain the route followed by the early trail-blazers."

Others describe the topography and environment in a similar way. After noting the views from the tops of the ridges nearby, the *Meyersdale Republic* cited "the Quemahoning Dam, one of the largest, and certainly of the most beautiful, artificial bodies of water in the State." The article continues to note the other historical items of interest in the area, including Fort Stonycreek and its linkage between Forts Bedford and Ligonier. "Looking on these autumn painted hills of maple, sumac, oak and pine, it requires no great stretch of the imagination to see the bright plaid kilts of Colonel Bouquet's Scotch Highlanders and hear the echoes of their bagpipes in the rustle of the wind among the leaves."[70] This represents the same combination of factors meant to elicit in the reader the same feelings of the sublime that the writer apparently felt: the convergence of history, the topography and the beauty of the natural surroundings, aided and abetted by the construction of an "artificial" body of water by a steel company, with no mention of the name of that company or, indeed, that the Que had been built by anything other than "artificial" means. Also at the mid-century mark, the Quemahoning Reservoir and Dam had been in existence long enough as a landmark that it would, along with other notable features

View across a northern stretch of the Quemahoning Reservoir. *Author's collection.*

of Somerset County, be an object of visual art. A local watercolor artist, Esau Motovich, publicly displayed his work at the Mountain Playhouse, near the Que. The exhibit showed paintings of local covered bridges and legacy one-room school buildings and featured the Quemahoning Dam.

Notes in various outlets would commemorate events of community interest occurring on Bethlehem property—picnics for retirees at Bethco Pines, for example. Even so, the outdoor areas were not entirely off-limits to the public, as other newspapers reported on hunting seasons and relative success of the archery whitetail deer season and waterfowl hunting. This can be thought of in several different ways, but through the review in this work in terms of the transition of a large technical system, we can see that the reservoirs and surrounding areas continued to be used for a variety of purposes other than industrial water. The mentions in the newspaper and the nature of the usage demonstrate, perhaps, a passive and shared understanding of that usage; the existence of the resource was taken as a given, along with the possibility of its shared usage. Indeed, further underscoring this point comes a hunting report from October 1955 that while the deer archery season had a record turnout in the area, there were no reported successful harvests; waterfowl hunting—at the Quemahoning—was said to be "fair." This is yet another example of public usage, reported in the newspaper, that appears to make things difficult to square with the notion that both Bethlehem and its predecessors actively excluded such uses.

The System Grows

In 1960, an important change to the system occurred—the addition of the Wilmore Dam and Reservoir. Wilmore Dam was not constructed by Cambria Steel, its successor Bethlehem or the captive water company, MWC. Instead, it had been built by the Pennsylvania Railroad in 1908. The construction of the dam and its usage emphasize the heavily intertwined and concentrated network of industrial undertakings during the peak of the heavy Industrial Revolution. The PRR had important depots and stops in the East Conemaugh area of Johnstown, near the massive Franklin works (constructed in 1898) of Cambria Steel. PRR itself built a 24-inch pipeline leading from Wilmore to the Conemaugh depot and the town of South Fork for the purpose of watering steam trains and for the use of its shops in those areas. It is, therefore, easy to see of Bethlehem's interest in acquiring the 800-foot-long dam when the PRR wanted to part with it, given how close the pipeline was to its own works, the congruence of the ongoing need for water for steelmaking and the price of $1.[71]

A technical report from 1979 describing the Wilmore Dam, quoted in its entirety, gives a good picture of that water since its acquisition nearly 20 years prior:

> *The Wilmore Dan and reservoir lies just east of the axis of the Wilmore Syncline, a north-northeast trending structure. The rock strata are members of the middle portion of the Conemaugh Group (Pennsylvanian Age). The rock strata consist of shales and sandstones and the bedding dips approximately 300 feet per mile to the northwest. The coal that crops out in the lower portion of the slopes is probably the Harlem coal seam. There are five major coal seams in the Conemaugh Group, none of which are minable. However, information obtained from the Division of Mine Subsidence Regulations, Bureau of Land Protection, Department of Environmental Resources, Commonwealth of Pennsylvania, and a map published by the*

> *same agency entitled, "Mine Map of Cambria County," dated January 1975, indicates that three coal seams of the underlying Allegheny Group have been mined up to the safety zone lines of the dam and reservoir. The available information indicates that the dam and reservoir have not been undermined. The mined coal seams are the B (Lower Kittanning), C (Upper Kittanning), and E (Upper Freeport) coal seams. The Upper Freeport coal seam, the highest of the three, is located approximately 400 feet below the ground surface. The steep slopes in the banks around the reservoir indicate that sandstone beds predominate, precluding large slides, although small rock falls probably could occur.*

The combination of factors here are profound, not the least of which is the mention of Pennsylvanian subperiod, dating from 300 million years ago, up through the geological awareness of coal, the harvesting of coal and the concatenation of bureaucracy holding an answer written once by an expert, reproduced by another one, at the request of a different expert in a different agency, all aligning with the possibility of safety and the interaction of humans and nature. In a sense, the construction of the Wilmore Dam

This page and opposite: Bethlehem Steel's reconstruction efforts at the Wilmore Dam in 1960. *Courtesy CSA.*

is a ratification of the natural underpinnings of the techno-natural system, much in the same way was the earth-filled core of the Que.

Physically speaking, the Wilmore Dam stretches for 800 feet from end to end and has a maximum height of 48 feet (which was augmented by Bethlehem after the 1960 purchase). The construction of the dam is not primarily an earthfill one but is instead composed of masonry consisting of fired clay with riprap and other features added by Bethlehem Steel (via the MWC) to bolster its strength and improve it. The construction, therefore, makes for a different appearance and a different set of performance parameters. While "only" around 200 feet shorter in length than the Quemahoning, the spillway capacity of Wilmore is a much smaller fraction of the Que. The nature of a masonry dam is such that being overtopped isn't the sign of danger that it would be for an earthfill dam, and the spillway construction for Wilmore is different as well. As shown later, however, the Wilmore Dam continues to cast perhaps the longest shadow of the three.

Even damage not caused by sabotage or other malfeasance can bring significant attention to the operation of the mill. In October 1961, the

The breast of the Wilmore Dam as Bethlehem Steel finished its refurbishment in 1960. *Courtesy CSA.*

The breast of the Wilmore Dam as Bethlehem Steel finished its refurbishment in 1960. *Courtesy CSA.*

main pipeline running from the Quemahoning to the Bethlehem facilities broke in downtown Johnstown. As mentioned in the engineering and design summary of the system, this is a 66-inch pipeline, and the notable instance of this for consideration of the true scale of the water is that the geyser spewing from the break was estimated to be shooting 100 feet into the air. For a leak in a 66-inch pipe to blow water 10 stories into the air requires a tremendous amount of pressure and volume. The news of the break made national headlines, certainly in the closest major daily in Pittsburgh but also the *Baltimore Evening Sun*, which also included a photo of the geyser and noted that in addition to water, the leak caused stones to be thrown high into the air as well.

The system was also used for drinking water provision over the decades. One prominent example occurred with the borough of Ebensburg—about 20 miles north of Johnstown and a few miles away from the Wilmore Reservoir—which began to acquire water when it appeared that its normal water source, Lake Rowena, was no longer adequate. There had been a drought in late 1963, such that it even affected the business of the Christmas tree growers in the area It also made for a slowdown at several Bethlehem steel plants in Pennsylvania and a series of rolling electrical grid cutbacks due to low water levels at several reservoirs. It had only been a few months before for news of many mudslides, snowslides, road closures and flooding conditions in Somerset County.

As time passed, the Que would be the setting for other events but no longer the news itself, such as in reports of burglaries at the Bethlehem-owned facilities and of unfortunate drownings and foul play, as occurred in 1973. It would remain the setting for public notices for hikes, such as with the Audubon Society, and for public awareness and cause rallies, such as a Bike-Hike.

The coming of the economic crises of the 1970s, Watergate, the Vietnam War and many other events of the time appeared to have permanently shifted the nation's self-perception, at least in some important ways.[72] It is, to use a pun, a signal of a watershed moment, moving the system and its engineering wonder to the background, and perhaps even consigning it to the past. The notions of fear of catastrophe, of course, continue, as do the politics and economic discussions of handling the system within the context of a world, region and city in a significant period of change. Still, the awareness of the system *as a resource for the area* gradually evolved as it became apparent that the presence of the steel industry itself was under pressure.

It is during this era that many perceived the United States as at its peak. The postwar years were decades of great prosperity but also periods of great social change. The last major labor actions in the steel industry and in Johnstown occurred during this time, but the first throes of significant and permanent economic downturn began. Pressure from foreign competition, escalating costs and largely uninspired strategy by leadership all contributed to the buckling that would occur in the 1970s. Plans were already underway in the 1960s to curtail steel production at the large but also largely inefficient Johnstown facilities, but these were tabled, most likely due to the fact that the then-head of Bethlehem Steel hailed from Johnstown. However, decisions postponed are sometimes left for circumstance to prioritize.

What goes up, as the saying goes, must come down, and the 1970s were unkind to the steel industry, its management and labor alike, and no areas more profoundly than in Johnstown itself. The major steel crash occurred in the early 1980s, the culmination of a series of events from the 1960s and 1970s; manufacturing technologies changed while global competition grew heated. Steel producers from Europe, Asia and South America were all competing by then, although mini-mills such as those pioneered by NuCor were the more significant competition, bringing more efficiency and lower costs to steelmaking.[73]

While the domestic steel industry had long been exerting itself politically, the politics of the times had shifted. Defensive trade measures to protect specific industries from foreign competition were not as in vogue as what they had been at one time; combining this with a general economic slowdown and other significant social issues in the 1970s, the will to impose such measures via political maneuvers had disappeared.[74] With the economic conditions of inflation, unemployment and product shortages, cleaving off special favors for an industry no longer such a dominant political and economic force, with large swaths of entrenched management doing very little of any strategic vision and labor unions seeming to be largely fixed on eternally elevating the living status of its membership, there seemed to be no good guys to pick in a play that featured only greed, perfidy and guilt.

The industry and its labor-centered enablers may have been able to hold off foreign competition, at least in some respects, but the technological and operational breakthroughs of the mini-mills could not be overcome. The combination of the two, against the backdrop of the "malaise," spelled doom for both labor and the companies. Tied together in mutual combat for the better part of a century, all the while increasing prosperity for themselves and each other, they fell into the same pit together, each side still wielding blades against the other and unable to relinquish their grip on power: economic, political, social. As such, they tied their fates together and, in so doing, also ensured their mutual demise and fell into the abyss as one.

The year 1977 is another etched in the common memory of those in the Johnstown area. Five years earlier, the region had escaped from the widespread flooding wrought by Hurricane Agnes, with the channelization of the area rivers that had been done after the 1936 flood thought to be responsible for the protection of the city and its people. They were wrong—again.

4

ENDS, AND THE BEGINNINGS OF TRANSFORMATION

1977–2000

On the night of July 19, 1977, a series of thunderstorms stalled over the Johnstown area, bringing more than a foot of rain in a 24-hour period by some estimates. One of the most recurrent memories of eyewitnesses that I've heard speak of that night was the lightning—it was constant, being in a world lit only by unrelenting strobe lights. As a consequence of the stalled thunderstorms and the rain, a number of dams in the area were overtopped and failed, killing 84 people. In addition to the rain itself, the dam failures flooded Johnstown with 130 million gallons. Clean-up took quite some time, with the digging out of rubble, the clearing of mud and the discovery and removal of the unfortunate victims, spread between Windber past Johnstown downstream to Armagh. The community of Tanneryville on the western edge of Johnstown and below the Laurel Run Dam was largely destroyed when that dam let go.[75]

The flood caused hundreds of millions of dollars in damage, much of which was inflicted on the Cambria facilities of Bethlehem Steel. At that point in time, the Johnstown operations were already in economic distress, as was Bethlehem Steel more generally and the domestic steel industry in its entirety. With the damage wrought by the 1977 flood, Bethlehem Steel management viewed it as a poor decision to attempt to restore the operations lost by the damage; instead, they cut more than 2,000 jobs. With this, of course, many could hear the death knell already ringing.[76]

The important aspect to keep in mind with this event was that while a couple of the six dams that failed were owned by Bethlehem, none of the dams within the current system did, including the Que. The Quemahoning Reservoir alone contains 12 billion gallons—such a failure would have likely killed thousands. News reports of the time, of course, had no little fodder to report on the death and destruction, including retrospectives on the 1889 calamity and the 1936 St. Patrick's Day flood. The *Latrobe Bulletin* contained quotes from David McCullough, the historian and writer whose 1968 work *The Johnstown Flood* set the standard for historical investigation into the 1889 event: "It's ironic that they put [Grandview] cemetery high in the hills after the floods. The dead were safer than those who survived it." Another eyewitness, on seeing the inundation of water, mud and wreckage in downtown Johnstown, described it as "the same as the 1936 flood'"

From Out of the Mud

As time passed, the Que, along with the other parts of the system, continued to be a site that exemplified this process of backgrounding itself. That is to say, the system was so well-established that it had begun to be taken as a given, as if it had always been there and always would be. It might even be said that it could have been considered a "natural" occurrence, though it was never thought to be anything other than an artificial lake. As it receded into the background of awareness, a constant barely considered, there also continued a parallel awareness of its value as some type of feature of the natural landscape. It was no longer understood to be an industrial water supply; it was, instead, an object of environmental veneration or, at least, something "out there in nature" that was recognized to be qualitatively different from a steel mill, building, automobile or coal mine. This becomes especially true when compared to the risks that presented themselves to the water, as in the case of the by-products of industrial usage in the area. Just as in the article from 1944 in the Pittsburgh press warning of the long-term dangers of acid mine drainage, any news of the Que seemed to take the form of a warning about possible despoilment, as in 1980, when a tractor trailer crashed nearby, leaking 14,000 pounds of residue such as iron dust and coal tar dye into the watershed and thereby exposing the reservoir to contamination. Still, the ties to industry and the heritage—and, perhaps most importantly, the memories and perceptions of the residents of the area—mean that economic development via heavy industry was still very much in the minds of many of the time, as during the construction of an Abex plant near the Quemahoning.

Now as a "quasi-natural" resource, a techno-natural system, the result of a co-construction of engineering and topography, the Que has also been the subject of discussion as a resource for power generation. In 1981, for example, the *Somerset Daily American* reported that American Hydro Power approached Bethlehem Steel management about the possibility of using the dam as a site from electricity generation. Why the project never went forward can be surmised; with Bethlehem and the rest of the steel industry struggling, it is likely that any further repurposing that would require capital or have an effect on share value would have been declined. There is also an indication that the proposal got far enough along that the electricity company filed for permits in late 1980. It is notable, however, that the article states that the area "is sometimes used for recreational purposes by Bethlehem officials." Of course, by now we know that while Bethlehem indeed used it for steelmaking and recreational purposes via Bethco Pines, there were plenty of public uses as well, whether officially countenanced by Bethlehem or not.[77] For example, at the same point in time, other announcements and various events continued, as in 1984, when the Quemahoning area was the site of the Journey for Sight, a fundraiser for the blind following a 10-mile bike, run, or walking route around the reservoir.[78]

Indeed, in the 1980s, after much of the contraction of the Bethlehem Johnstown facilities, there is an indication that some support had been growing for the conversion of the Que reservoir area into a public or state park, supported by a local hunting and shooting club, the Jenner Community Sportsmen. Even prior to this, considerations had begun, with the filing for grants to study the Stonycreek River to clean and better manage the watershed. The Que never became a state park, either, though, as we will see, eventually the usage of it unfolded into something that could look like it. Throughout the 1980s, it was evident that Bethlehem Steel was looking to eliminate its cost centers as it continued to contract operations globally and especially in Johnstown. Still, in the early part of the 1980s, Bethlehem's intentions did not seem entirely clear. Acting on news that the company was going to divest nonproduction facilities, Somerset County officials sought to bid on Bethco Pines, which was soon rejected with a statement that Bethlehem was not interested in selling. There was also a proposal to release Bethco Pines and a restaurant, golf and tennis facility named the Ye Olde Country Club in suburban Johnstown to the ownership of employees who

A deep coal mine near Elton, Pennsylvania, owned by Bethlehem Steel subsidiary Bethenergy, in the 1970s, as the company entered its final decline. *Courtesy LOC.*

The Finest Coal Miners In The USA.
COAL
you can feel great
and still have
HIGH BLOOD
PRESSURE!
have it checked

sought to retain possession of the facilities. Some transactions of this kind occurred as Bethlehem wound up the 1980s, eventually selling Bethco Pines to employees, with an interesting consequence of this transaction being the wind-up of a Scout camp, Roaring Run, which had been on the list of bidders for the property from Bethlehem. It was not to be, however, and the camp—also located in Boswell near the Que—closed in 1986.

In 1983, a concerning instance of environmental degradation occurred. At this point, the watershed was still under the control of Bethlehem Steel and MWC, but as with anything in this day, there are a number of easements and rights of way, including to utility and other extraction companies. In one case, it involved an oil spill in the Wilmore watershed, for which Gulf Oil and its contractors were responsible and engaged in clean-up. Correspondence between Bethlehem Steel and Gulf Oil indicated a plan of containment, clean-up, and monitoring performed by Gulf while Bethlehem provided hydrology information and other related data. Not much can be inferred from the correspondence, as it was cordial and businesslike, rather than rancorous or readying for litigation. It was, instead, clear that Gulf understood the need to clean up after the spill, and Bethlehem's expectations were such that it would be done and assisted by giving the data needed *and* the approval of the actions of the clean-up itself.

Coal mine wash house near Revloc, Pennsylvania. Coal mining in the region is at least as much a part of the identity of the people and economy as steelmaking. *Courtesy LOC.*

The discourse surrounding the usage of the Quemahoning area, along with the other reservoirs, slowly grew through the 1980s and continued through the 1990s. As the once-towering shadow of Bethlehem Steel began to fade, new possibilities came to light, bringing new contrast to the contours of social reality in the area. The smoke pouring from the stacks and the coal issuing forth from the ground slowed and then, at least in the case of steel production, stopped entirely. The new possibilities, with the literal and figurative clearing of the air, meant that there were far fewer jobs, an increasing exodus of the younger generation from the area and a loss of relevance to the economic health not only of the area or even the region but also from the world. An identity based in industrial might faded, leaving behind empty homes, fewer businesses and smaller schools. Yet at the same time, there would be a new view on a clearer sky, trees no longer denuded of leaves and clear waters pushing against the surge of the orange tide of acid mine drainage.[79]

While we have seen early examples of dual usage of the area as a large technical system, virtually since its inception (and arguably, the part of the "dual" use as an outdoor resource preexists the eventual and current use as an industrial water supply), the system itself was relatively stable from an economic, social and property rights standpoint from its inception to the time that Bethlehem began to totter. Through the review of this history, we've seen significant and growing reportage on the environmental and historical aspects of the system through the 1970s, perhaps as a way of looking backward in time before steel or perhaps as it sat—independent of any consideration of the industry that built it. But the ownership of the system and what to do with it had been an active, public question since the rapid decline of Bethlehem in the closing years of the 1970s and throughout the 1980s. Various uses were ventured as possibilities, but many of these were undoubtedly drawing clouds of conflict. In 1987, for example, there was a squabble over some negotiations that had been taking place for the sale of the Quemahoning to the Greater Johnstown Water Authority for $4.5 million.

We come back to the question now that the era of stability has ended—what to do with a large technical system when its governance and ownership is in question or even dying? What to do with billions of gallons of water stored *uphill*?

One option appeared to be to outsource ownership to the state by placing a prison on the property. A letter to the editor of the *Somerset Daily American* in January 1991 asserted that some made an argument "to our commissioners and State Representatives the people of Somerset county want a prison even if it is located at the Quemahoning Dam site. Remember, the people in this Quemahoning site have stated for the last nine months that we *do not want* the prison located here which has been evidenced by the thousands of signatures" (emphasis added). Letters also were posted earlier, in 1990. The site was not and has not yet been chosen to site such a facility; indeed, under the current ownership and usage, that appears to be an outcome well-avoided and something that is currently unlikely to happen.

The idea of maintaining the water as it would eventually become, in the form of the CSA, took years to develop. As early as 1988, commissioners from Cambria and Somerset Counties were meeting to consider the possibility of a purchase of the reservoirs from Bethlehem. This was complicated by the fact that Somerset County at that point was suffering from a necessity to cut spending. Several usage and pricing considerations were underway as well, settling on an amount of $4.5 million, short of what the actual bid would become a decade-plus later (although in 1988, the entirety of the system that became the CSA was not the subject). By 1990, a candidate for the state house of representatives issued a petition with more than 1,000 signatures to Somerset County officials to convert the Quemahoning to a public fishing area. A year later, the candidates openly discussed the possibility of such a purchase.

One of the realities striking the area in the 1990s was the continued outcome of the contraction of the steel industry that began in the 1970s. In 1997, for example, Bethlehem Steel was essentially insolvent and would be bankrupt in a handful more years. All of Bethlehem's steelmaking activities in Johnstown ceased in 1992, and in 1995, it closed manufacturing in its namesake city of Bethlehem, Pennsylvania. In 1997, it also wound up its maritime business activities; in short, the company had been in the mode of a fire sale for the better part of a decade. The Que was still under Bethlehem ownership, and it appeared the company was growing increasingly desperate to offload the reservoir.

Six months after that, Bethlehem was still the owner of record for the Que and actually the enabler of a glimpse of the drowned history of the Que. The company drew water levels down in the reservoir to repair the spillway, and when the water levels got low enough, the ruins of the old town of Stanton's Mill could be seen. It was an unusual sight—a ghost town that hadn't been glimpsed in nearly a century.

> *Most of the buildings were torn down when the Quemahoning Dam was built in 1911...but stone was left alone. At least 20 foundations can be*

> *seen...*[with] *30 more...still under water along an old creek bed at the bottom of the reservoir.... The water had to be lowered for modifications to the spillway...said Gary Graham, a spokesman for its owner, Bethlehem Steel Corp. Later this year, the reservoir will be filled again.*[80]

Where Bethlehem found the funds to repair the dam is anyone's guess, given the state of the company at that point in time. Still, as Vinsel and Russel discuss in *The Innovation Delusion*, the necessity of maintenance inevitably becomes apparent. Sometimes, as in the case of floods or other disasters, it comes too late; but even as Bethlehem was on its last legs, it was still maintaining a dam that it no longer operated or, indeed, could likely afford to operate, since the steel mills the reservoir had supplied closed all operations years before and all major operations nearly two decades prior.

There might even be several points made, or at least considered, about the prominence of the Que, Wilmore and Hinckston during the 1990s. Other than mention of a draw-down and the occasional notice of a public meeting or perhaps the rare public interest story of being the site of a cause-awareness hike, it seemed to be increasingly seen as an infrastructural burden. Replacing primacy would be, again as alluded to by Vinsel and Russel, a grudging acceptance and with that a begrudging eye toward them. Fatalism replaces dynamism.

Perhaps when change ceases to occur in the places it had been ordinarily seen, it has to happen somewhere else. One clue of such a change came in late spring of 1998 in the *Somerset Daily American:*

> *NOTICE: PLEASE BE ADVISED that the Somerset County Board of Commissioners will hold a Special Meeting with the Cambria County Board of Commissioners on Thursday, June 4, 1998, immediately following the Cambria County Commissioners' Meeting.... The purpose of this meeting is to discuss making an offer to purchase the Quemahoning Dam.*

This humble public notice marks an important turning point, the signal that the direction that would eventually be cast was taking place.[81] The industrial development organizations in the region strongly supported the purchase of the system by the counties, despite the objections of organizations such as the Greater Johnstown Water Authority. A local columnist noted that "everyone wants to get in on the reservoir purchase," and the background events at the same time showed a series of meetings by the county commissioners

Quemahoning Reservoir looking north. *Author's collection.*

considering adjustments to their bid. At last, the fading steel giant had found a bidder for its infrastructural burdens, and that buyer would be, essentially, the public. By November, it was a done deal.

Following this comes a brief summary from the *Indiana Gazette* in November of that same year, which also encapsulates some of the most important areas of this change:

> *Bethlehem Steel Co. sold a water company in southwestern Pennsylvania, which includes five dams and reservoirs, to Cambria and Somerset counties, which hope to use the water to spur industries and recreational development....The watershed that fills the five dams crosses political borders, a geographical fact that persuaded the counties to approach Bethlehem as partners.* Mindful of the counties' development plans and mutual water needs, Bethlehem apparently turned down more money from private developers to accept Somerset's and Cambria's joint bid of $6.25 million. *"We believe that Cambria and Somerset county officials presented the most responsible proposal for the future of the two-county area and that future development of the region*

> *will benefit immensely under their ownership and leadership," said Curtis Barnette, Bethlehem's chairman and chief executive officer.* Bethlehem also agreed to pay $1 million for repairs to the spillway at the Quemahoning Dam. [emphasis added]

There are several other items of note in this piece. First, the degree of competition for the ownership of the dam stands out. The joint bid by the counties exceeded that of another public authority, the Greater Johnstown Water Authority, but apparently was not the absolute highest bid in economic terms, which would have likely come solely from private land developers who would have almost certainly found great residential demand for home sites in the area, or even other commercial opportunities that would be as interested in, or even primarily interested in, the aesthetics and natural beauty of the area (albeit with artificial dams.) Nonetheless, Bethlehem eschewed that route, choosing from the two competing proposals. More insight from that piece discloses what might have been a bit of tension, or even scorn and disputation between the two bidding parties, again from the *Gazette*:

> *Board members of the Greater Johnstown Water Authority had discussed trying to block the sale if its bid was unsuccessful. Now that a deal has been struck, two authority members said it should go forward. "We're very disappointed we didn't get it. But now that the die is cast, my gut feeling is we would do anything to help the county commissioners to run it effectively and efficiently," board member Charles Glass said. The authority pulls 8 million gallons a day from the Quemahoning to supply water to 17 municipalities and its officials were concerned that development around the dams could pollute the water. But Somerset County Commissioner*[s] *Dave Mankamyer and Soissons said they would make sure that any development would preserve drinking water supplies.*

What happened was this: in late 1998, a committee of equal representation from both Somerset and Cambria Counties, represented by commissioners and solicitors from each entity, met to negotiate and then finalize the agreement with Bethlehem for the sale of MWC. This occurred after the bid submitted by both counties had been reviewed by Bethlehem, and at that point in time, the actual form of the organization that would become the MWC successor was not yet fully fleshed out. The Cambria-Somerset Authority was the next step after the consummation of the sale of the system.[82] There was somewhat of a delay that perhaps had not been fully anticipated, which necessitated further conversation among the parties. The agreement was also pending approval by the Pennsylvania Public Utilities Commission, and there had already been entreaties from neighboring municipal water authorities about the possibility of purchasing drinking water from the new entity. Things were, in a sense, moving both quickly and slowly as the consummation of the sale and its ramifications loomed. Still, some disputes remained to clean up, including lingering rate increases for a half-dozen or more industrial water customers of the MWC, prior to the actual transfer of ownership to the new entity.

From Steel Malaise to Green Ecology

The agreement was the result of an extensive negotiation and bidding process, but other points to be considered are some of the other stakeholders present in a regulatory, enablement and other ownership forms that were included in the administrative structure of this LTS. The deal would only be effective after its approval, for example, by Pennsylvania's Public Utility Commission. In fact, there would be litigation stemming from the regulatory structure, when the CSA sued the Greater Johnstown Water Authority for its protest against the Bethlehem conveyance of the system to the CSA. Obviously, the system eventually won approval in one way or the other, but this is an illustration of the nuanced complexity that can occur within a highly detailed regulatory structure and the underlying legal system, which, fortunately, accounts for some degree of due process and justiciability. The counties would also have to create another administrative authority for the operation of the system, along with execution of public finances to marshal the resources to make the purchase. They then issued bonds to finance it, with the expectation that the sale of the water would rapidly erase the debt and repay the bonds. There was, in essence, a significant layering of permissions and checks on the entirety of the deal; is it worth the inevitable complications that would arise then, nearly a century since when Cambria Steel Company first made its moves to create the system.[83] Amid all of the official actions to conduct the negotiations, bidding and purchase, authorities and other groups affiliated with and possessing interests—and power to solicit and gather information and opinions from interested members of the public—also sought, received and distributed input. The Southern Alleghenies Conservancy was one such organization mentioned, among others, as working to gather information pertaining to the possibilities of recreational plans for the system throughout 1998 and 1999. In addition, in the late 1990s, a host of activities began, devoted toward fisheries improvement, acid mine damage remediation and soil and water reclamation in the Quemahoning area and throughout Somerset County. Amid this backdrop, Somerset and Cambria County officials negotiated with other potential bidders, asking the interested parties not to compete with the counties' efforts to purchase the system, noting that a bidding war helps no one and escalates the eventual price for all.

Given the economics and the contrast between public and private sector incentives, one might conceivably ask the question: Why did Bethlehem, a company with rapidly dwindling resources (indeed, if any at all), pick the path that it did? Surely an answer based only on finance and economics would conclude that Bethlehem had made an error by failing to maximize its income on the deal. This decision is not, however, the proximate cause of the Bethlehem Steel bankruptcy; it is not even remotely related to it. Instead, the combination of labor actions, management decisions, global markets and global changes in politics and demand over the span of decades is what situated that company—along with several others—on the brink of extinction by the turn of the millennium. It elected not to sell to any private parties, even at likely greater economic reward. Perhaps they remembered the dangers of dams when maintained by private and unaccountable landowners, such as with the disastrous 1889 flood, and elected instead to go with a public authority that would, in theory, have greater accountability and transparency leading up to steps that would be taken prior to any type of catastrophic failure.

Having ruled out a private ownership possibility for the system, Bethlehem made a decision more consistent with finance and economics—they selected the highest bidder from the remaining public options. In considering the words used by all the parties—including Bethlehem—there may have been still other many valid reasons for the selection of the counties' bid over that of the Johnstown water authority, but the fact remains that the counties' bid was $250,000 higher. In the end, it culminated in a combination of public-mindedness and financial common sense.[84]

The opening usage of the Quemahoning and the related system occasioned a sequence of outreach and fact-gathering efforts, including solicitation of ideas from the general public. A range of ideas emerged, from the restoration of Bethco Pines and its swimming facilities, to dinner cruises, boating and increased opportunities for hunting and fishing. One

From the western shore of the Wilmore Reservoir, viewing to the east. *Author's collection.*

of the areas of balance that occurred early in the process as well was a reminder of due regard for balance and public usage; to the extent that the facilities and properties were held in private by Bethlehem Steel, the notion that the resources were now public ones was a prominent refrain. Equitable access, therefore, and appropriate budgeting for maintenance were all mentioned as desirable outcomes—certainly for the Que but also for Hinckston and Wilmore. The public input sessions held at various locations throughout both counties elicited a recognition of a strong desire for boating, fishing and hunting opportunities. There were also a number of concerns, many of which have been borne out, such as vandalism and destruction of property through overuse or wanton use of things such as ATVs. Water quality was also thought to be of concern, but perhaps most interestingly, the need for ongoing maintenance was not apparently considered, or at least mentioned, as one of the countervailing necessities of the management of such a system.

A ramification, perhaps unanticipated, perhaps not, with the ownership of the waters being a public resource is that they would then become political and economic leverage, even of a sort that had not been as readily seen when owned by Bethlehem or its predecessors. Instead of

This page: Views of the Quemahoning Reservoir and the natural system surrounding it. Note the clarity of the water. *Author's collection.*

an economic input for production purposes, the infrastructural system becomes an object of power, a political talisman packaged and marketed in the consumer preference process of local elections. One Somerset County commissioner at the time, when campaigning for reelection, stated, in response to a question about his plan for the betterment of economic conditions: "I have taken the first step to increase economic development by voting to purchase the Quemahoning Dam. This purchase will provide the water needed to create jobs in our county. The sale of water from the entire Manufacturers Water System will generate enough revenue to pay for the purchase. This project will provide the infrastructure needed to sustain our region into the next century without increasing any tax burden on the residents." Even as years passed, the exemplification of the outdoor recreational usage of the system continued to be a central point for election and reelection for office holders. The sale would be effected in late August 2000. The Utilities Commission had rendered its approval a few weeks earlier, and a few other officials were named as being instrumental to the completion of the sale. In addition to the combined efforts of the county commissioners working together, the sale had the support of the powerful local member of Congress Jack Murtha and then-Governor of Pennsylvania Tom Ridge, who would later become the inaugural secretary of the Department of Homeland Security under the George W. Bush administration. State-level elected politicians also showed support, along with the Pennsylvania DCNR secretary, John Oliver.

The article reporting this closed by mentioning that the latest business report of the MWC in 1999 was the total sale of water for the year being 11.5 billion gallons for $1.6 million. They reported that the current usage of the water system was 30 million gallons of water per day, nowhere near its total supply capacity. It was also listed as having 14 customers at the time, including the drinking water companies for Johnstown and neighboring Windber Borough.

THE NEW CENTURY

The year 2000 created a zeitgeist around the world, from the "end of history" as declared by Francis Fukuyama a few years before to some species of panic associated with the change in the year to two-digit computer date systems (the Y2K problem) and everything in between. In

Sign at the entrance of Wilmore Reservoir; it is important to note the consistent and constant presence of a variety of stakeholders in the policies and execution of the transition of the water supply to a public resource. *Author's collection.*

this area, and for this system, it was the dawn of a conscious change in the meaning and purpose of the Que and its related bodies of water from "industrial supply" to an area with viable conservation status. It's no longer a "site" for site-seeing, for awareness hikes or annual runs,[85] for fishing or hunting reports or to be labeled as the private retreat for a wealthy few within Bethlehem Steel.[86] Instead, it is a site *of* the natural world, an all-season showcase in subtle splendor in all four seasons. Perhaps it was the coming of the turn of the millennium or some other such movement that had been in the works for a while, this pivot toward the environment. The CSA system has been at the center of all of it, however, and has been for a considerable length of time, including a writer in 2001 noting the return of bats and other insectivores to the Que and the importance of the food web to the ecology not just of the region but of the environment generally. That same year, 2001, is the year that fishing was not only tolerated but also welcomed.

The Bethlehem decision to award the system to what would become the Cambria-Somerset Authority can likely be seen as one of the most consequential decisions for the heritage, history and environmental health of the area of the closing days of the 20th century and riding into the first decades of the 21st. The power flowing, as it were, from the last gasp of the legacy steel industry to consign an integral part of its own infrastructure to an intergovernmental entity transformed into the momentum that focuses on environmental stewardship. An artificial dam and reservoir, therefore, becomes the channel of power *for* the natural world, the transformation of the industrial system into a system trained toward sustainability.

Another example around the turn of the century: in July 2000, a piece in the *Somerset Daily American* explored the new efflorescence of awareness of the possibilities the area's waters, terming it a "River Revival." The dual nature of the senses of the word *revival* seem fitting—a revival in terms of a restoration of environmental promise but also nearing a second meaning in terms of religiosity. The piece describes the Conemaugh Valley Conservancy and its local role in environmental improvement projects. In 2000, the Conemaugh-Kiskimenitas River was Pennsylvania's River of the Year, a dramatic turn from its decades of being the recipient of significant acid mine drainage. The article describes the course of the rivers and the pattern of the watershed and also discusses some of the major sources of pollution that hinder the river's full recovery and the Conservancy's efforts at fostering recreational usage of the area, along with incorporation of the Stonycreek River and, therefore, the Quemahoning Reservoir. It was, according to a source in that article, "the greatest conservation effort ever enacted in the history of this region." This then sets the stage for the third act; steel is gone, coal is much diminished and far different and 70 percent of the population has disappeared. What remains of the rivers, reservoirs and dams?

5

MODERN HISTORY, MODERN MAINTENANCE AND MODERN CONSERVATION

2000–TODAY

Around the middle of August 2021, in the warm waters off the western coast of Africa, a mild tropical disturbance formed and eased westward. Over the next several days, this disturbance grew in organization and appearance. By the time it reached the Windward Islands, riding the quicker east–west currents and trade winds, the system had developed into a full tropical cyclone categorized by the United States' National Hurricane Center. The NHC categorized the system as "highly likely" to form a hurricane as it took a distinct turn northwest through the Gulf of Mexico, heading for a direct impact with Louisiana. It was assigned the moniker "Tropical Depression Nine" as it neared Jamaica, and humid air, warm waters and favorable winds all led to the strengthening of the storm as it neared the Caymans. It made its initial and transitory landfall in Cuba with 80-mile-per-hour winds, then gathered strength and picked up speed as it headed north-northwest.

Hurricane Ida hit the southern coast of the United States at the Mississippi Delta on August 29, having gained 70 miles per hour *more* than what it had had just hours before, generating sustained winds over 150 miles per hour. It tied two previous hurricanes for having the highest sustained winds and was second only to the 2006 Hurricane Katrina for the amount of atmospheric pressure drop. As it came ashore, Ida held onto Category 4 status for hours as it cruised northward and inland, preying on the flat landscape of Louisiana. It reduced to a Category 3 and maintained that for several more hours, then de-energized to a tropical depression as it surged north-northeast.

It reached central Appalachia on August 30 and continued its trajectory. No longer a hurricane, Ida had become what was called an "extratropical" event; meaning that it was outside of the tropics but still contained potential for high levels of danger for those in its path. Ida spawned tornados as it moved through Appalachia and to the Mid-Atlantic, bringing power outages to millions along the way, with tornado activity recorded in southeastern Pennsylvania and Annapolis, Maryland.

On September 1, the Johnstown-Somerset area came under threat from Ida. The storm brought high winds and prodigious quantities of rain. The topography of the area, of course, once again came into play and sharp relief when the National Weather Service issued the following bulletin:

> *FLASH FLOOD EMERGENCY FOR TOWNS AND CITIES IMMEDIATELY BELOW WILMORE DAM ON THE NORTH BRANCH LITTLE CONEMAUGH RIVER…*
> *The National Weather Service in State College has issued a*
> - *Flash Flood warning for…*
>   *A Dam Break on the North Branch Little Conemaugh River below Wilmore Dam in…*
>   *Southwestern Cambria County in central Pennsylvania…*
>   *Until 700 PM EDT*
> - *At 109 PM EDT, County Dispatch reported an uncontrolled release from the Wilmore Dam causing Flash Flooding downstream on the North Branch Little Conemaugh River.*
> - *This is a FLASH FLOOD EMERGENCY for towns and cities immediately below the Wilmore Dam on the North Branch Little Conemaugh River. THIS is a PARTICULARLY DANGEROUS SITUATION. SEEK HIGHER GROUND NOW!*

By 2021, there had never before been faster communications of emergency information; there was, at the same time, never a bigger risk than what there had been before with regard to amplification of falsehoods and rumors. For there had been no dam break at Wilmore.

Also by then, CSA had determined that the higher water levels justified the evacuation of the people below the breast of the dam, in the town of Wilmore. After a couple of hours, emergency responders noted that the spillways were venting water but the dam itself was intact. That evening, residents were allowed to return to their homes, with the precautionary evacuations then canceled. Social media reports largely drove the frenzy. Subsequent

Wilmore Dam breast, from the downstream, looking north-northeast. *Author's collection.*

evaluations suggested that the dam had never been at risk of failure and that the panic had spread, like a 21st-century, social media gasoline fire, through the sloppy use of words, panic and old-fashioned human nature driven to drama.[87] Some reports noted the evacuation of "more than 40,000 people" in the Johnstown area. As noted earlier, Johnstown itself has a population of only 18,000; for 40,000 to be evacuated, it would have been about a third of the entire population of Cambria County. In any event, social media disinformation handily outweighs legitimate knowledge in many arenas.

Were the lessons of the breathless reports from 1936 not heeded? Perhaps—but there were also catastrophic floods in 1977. Even as there had been reports of the failure of the Quemahoning and Wilmore Dams in 1936, the cataclysmic failure of six other area dams in 1977 cost many more lives than the flood in 1936. The National Weather Service shouldered the blame for erroneous reporting, and of course, the local newspapers, much like the Canonsburg outlet in 1936, congratulated themselves (perhaps rightly) on not falling prey to the onslaught of shrieking rumor. From the *Johnstown Tribune-Democrat* on September 3, 2021: "The Tribune-Democrat reported only verified reports from professionals positioned to know—and avoided unfounded rumors, even those called or sent into our newsroom."

Wilmore Dam breast, upstream, looking eastward across its breadth. *Author's collection.*

Wilmore Reservoir, at the northern end, looking northward. Notable here is water clarity, the growth of aquatic plants and riparian forest. *Author's collection.*

There had been flash-flood damage, however—that much was true. The storm also spawned tornados all over the region, predominantly around the Chesapeake Bay. However, the compounding effect of panic extracts much interest even when there is not enough principle to pay.

This is not to say that Wilmore Borough was without damage. Many basements had flooded, sometimes to the point of ruining the home with loss of appliances and equipment but also destroying irreplaceable items such as family heirlooms and keepsakes. "We lost everything" was a common refrain heard from residents who experienced flooding that early September. One citizen questioned why the water level within the reservoir hadn't been lowered in anticipation of the storm, perceiving that the Little Conemaugh would not have swollen quite so much or as quickly. It is likely, however, that the time it would take to even partially drain the reservoir would have had to have been planned at least days in advance, and a controlled release takes time—precisely the situation that occurred anyway. And as far as the optics of the release of water pitching forth over the center of the dam, the inspecting engineer averred that the spillways functioned as they had been intended and designed, and while the imagery might be quite fearsome, the dam performed well during the heavy event.

Still the Flood City?

Coughenour and colleagues, cited earlier in this book, also analyzed some of the findings from the 2021 event and Wilmore Dam. It draws attention to a report that the dam was rated poorly by the Pennsylvania DEP. The authors noted that while that was disputed, the dam also unquestionably survived the major events of 1936 and 1977, with the 2021 spillway situation an added data point to the dam's century-plus-long history. Earlier reports on the engineering of the dam emphasize the masonry structure of the edifice. The overtopping of the dam in case of heavy rain or storm runoff is not the signature death knell that it would be for an earthfill dam. Importantly as well, Coughenour and colleagues note that the events in 2021 were not the significant level of precipitation that fell in 1977.

Coughenour offers several important conclusions for consideration; one, as many or even most researchers do, they note that there is a need for greater study and modeling of the risk of the area in its totality. And while most researchers do call for more study, irrespective of the field, it seems to be especially prudent in an area with this history and this degree of potential risk. An event not having happened is hardly evidence for it never being able *to* happen. An ancillary note that the authors make emphasizes this, as the latest decades of the 21st century have been the absolute wettest on record for the entire region. Climate change can, therefore, play an important role in increasing that same, already-known hazard.

Whether the situation could be replicated in an area with less history of catastrophic flooding is truly impossible to answer; there are few places on the planet with the flood experience of Johnstown, Pennsylvania, at least in comparatively modern times. Indeed, few areas have allied themselves to catastrophic events in such a way so as to make them become part of the identity of the community. One of Johnstown's more prominent nicknames is the "Flood City." This is an interesting enough tagline for tourists, perhaps, but also one that has a dark reflection on the minds of many who live in the area, brought up in the shadows of one of the nation's great catastrophes of the 19th century and then enduring two more episodes. The area has not yet grown exhausted with shouts of fire in a crowded theater.

The bottom line is that words are important, and terminology matters. The essential miscommunication that occurred with Wilmore and, to a lesser extent, Hinckston Run, was that specific terms of art were used to describe both the dams and the weather, along with the descriptors for potential effects when the two combined. While the 1936 panic regarding the failure of the Que was a broad, simple and visceral rumor, the description of the events in the wake of Hurricane Ida was different. Terminology such as *breach* represents one problem and the way the dams are described as *high-hazard* another. This is not necessarily because of their construction or any type of flaw; indeed, the Hinckston Run Dam in particular is unlikely to breach under any set of conceivable circumstances given the amount of slag backfill that the Bethlehem plant deposited there for decades. Instead, the term *high-hazard* means that a failure could cause a catastrophic loss of life, with hundreds or even thousands of people in the path of the water that would be released in the event of a failure. The systems of inspection and review by the Pennsylvania government keeps a list of dams within the state for some period of time; the 1889 Johnstown flood was one of the major reasons why the index and inspection program apparently began.[88] With these as more recent events, we should look back to the start of the century for the history of the CSA itself, its entwinement in the community and its ongoing mission and challenges.

One might safely conclude that Bethlehem's ultimate decision to sell to the counties was largely public-spirited. Clearly the company was pursuing a strategy of radical retrenchment and cutbacks. As with so many things in considering the actions and the realities of the large technical system in a web of power, the term *public* takes on more than one meaning. With the inception of the Cambria-Somerset Authority to manage the system, the web of stakeholders that had become far more visible during the dealmaking process for the counties' purchase of MWC began to grow and multiply. It would lead to inevitable questions about maintenance, upkeep and the eternal query regarding catastrophe. Planned million-dollar improvements were in the works, including upgrades at the Que, operations at the Border Dam and pipeline repairs, for a total of $8 million. A state grant was also in play for $1.5 million, necessitating certain limitations on what the CSA would be able to do to raise money to finance the maintenance and improvement activities—it had to keep its usage fees low, for example, such that it would remain qualified for the grant. Also considered were increases in fees and, most strikingly, the possibility of strip mining for coal along one part of the Que reservoir.[89]

Aquatic and riparian life surrounding and within the Quemahoning Reservoir, in addition to supporting multiple outdoor activities, including hunting, fishing and boating. *Author's collection.*

Now as well, "maintenance" was at the forefront of the public agenda, no longer hidden as a set of lines in a massive budget of a steel company. It became so public, in fact, that the entire maintenance strategy was a part of meeting agendas published in newspapers, such as in an open request by CSA for bids on improvements to the Que on January 16, 2002.[90] Following a sealed-bid process, the entire request is worth reproducing, bearing in mind both the technical specifications and the network of interests that accompany them, from the *Somerset Daily American* in January 2002:

> *Work items for this project include +- 22,000* [cubic yards] *of earthwork and placement of precast concrete jumbo blocks necessary to raise the impoundment structure 5.5 feet to comply with* [Pennsylvania Department of Environmental Protection] *Dam Safety criteria for containment of probable maximum flood event, reconstruction of a portion of Quemahoning Dam Road, and placement of associated erosion resistant revetments near the dam spillway area.*

The notice also advised potentially interested bidders about the cost for the plans ($70) and the different locations where the plans might be viewed: Pittsburgh, Altoona and Monroeville. Also on the agenda was an in-person visit to the site. By September of that year, the project to increase the height of the Quemahoning dam—the first structural change to the actual dam itself in four decades or so—was underway.

The Infrastructure of Perception: Shaping Usage

As the handover evolved in the late 1990s and early 2000s, along with the explorations for different types of usage to be associated with the resource, the importance of partners with state agencies started to grow as well. The Pennsylvania Fish and Boat Commission and the Pennsylvania Game Commission were both involved in planning resource usage for angling and hunting, respectively. The 2002 wildlife management plan recommended focusing on habitat improvement and improvement of small game hunting opportunities, such as upland bird hunting for grouse and pheasant. The complex history of the area, with a mix of farming and forest and the riparian nature of the system, made for an interesting and varied set of management possibilities.

An ongoing environmental touchstone for the CSA techno-natural system is its role as riparian forest. The concept emphasized by CSA involved practices and realizations contained in an article by Matt Ehrhart on the management of riparian ecosystems and edges along a watershed. The wildlife management plan included the planting of different plant species to facilitate the betterment and upkeep of habitat; the riparian forest attributes include the cover for the stabilization of soil, anchoring nutrients in the soil and facilitating the development of invertebrates that serve as an important component in the food web and the ongoing viability of biodiversity in the system. This has knock-on effects throughout the locality, state and region.

The placement of the system in the network of regulations of public notices and multiple stakeholder deliberations, situates it amid different (even if largely consistent) visions of what the waters might mean to the public. The same would be applicable to other areas and initiatives which, while not directly a part of the CSA system itself, were related in the sense of being nearby watersheds or having a role to play in such multiparty partnerships. Just a few days prior to 9/11, U.S. Congressman John Murtha noted the efforts of and assistance by the U.S. Army Corps of Engineers for a related project improving the water pipeline from the Quemahoning.

Drinking water is critical to all of us, a necessity of life itself. Much of the politics in the western part of the United States has centered on, sometimes explicitly and sometimes not, on the availability of water. Availability of water is always a source of politics, however, even in areas where water shortages are not currently a real risk. The move of the Que, Wilmore and Hinckston to the public weal just pushed a bit more of those politics to the forefront; while the MWC quietly sold water to boroughs and systems in the area while owned by Bethlehem, now it seemed to be seen—or at least portrayed—as a different entity entirely, subject to the possibilities of mergers with other systems or seen as the chiseling newcomer seeking to deprive municipalities of their own self-sufficient water supplies. The construction of a water pipeline from the Quemahoning to the Somerset area to serve development of industrial parks was heralded as a necessary and highly desirable part of the economic plan and infrastructure for new industry. Communities and policymakers strongly supported the building of a water supply pipeline from the Que to support economic growth and to make the water supply overall more reliable. A letter of intent to purchase 900,000 gallons of water daily by Somerset Borough in 2004 also heralded future commitments and reaffirmed the essential premise of the CSA in its early days. Economic forecasts for the expansion included security for existing jobs and the potential for 2,700 new jobs, new housing and fire safety.

The politics of rate increases suddenly take on a new form when subject to a deliberative electoral and administrative process; the hot takes in newspaper articles about sudden rate increases may generate good headlines, but they also tend to generate far more heat than light. The improvements and necessary upkeep of quiet, simple and enormous machines like the Quemahoning Dam are an irrefutable fact of their existence. Later in 2002, after the debates about mergers and the public announcements, the CSA went forward with its additions to the Que, with engineers calculating that the new standards from the Pennsylvania government and the Que's modifications to comply with them meant that the "dam will be able to handle floods from 20 inches of rain in six hours." That is one half of the politics: handling compliance with changing government standards for infrastructure and at least arguably in the camp of public safety. The other half of the politics stems from environmental concerns or ecological occurrences—handling drought. Another CSA capital expenditure early in its history was the construction of new pipelines in the interest of water access for drought challenges to drinking water availability and for fire protection.[91] This is water and water access at the fundamental infrastructural level. Indeed, CSA had certainly felt its stride by 2005, with a number of projects for Hinckston and the Quemahoning open for public bidding and Earth Day celebrations to be held at the former Bethco Pines facilities. Other hooks, so to speak, came from the angling community, when the broader group of interests encompassed by the Stonycreek River watershed focused on the first-ever fishing tournament to be held at the Que. Some years later, a column by Dave Hurst noted that the fishing tournament and other recreative aspects in the area would at one time have been unimaginable and were now not only broadly accepted but also embraced (except for, as we've noted, the fishing derbies that occurred at Hinckston in the 1910s). Hurst noted several important reversals, or at least contrasts, from the days when the water was under the direct control of Cambria and later Bethlehem—namely, public access, forest management around the waters and other items that are of considerable improvement to quality of life in the area.

The political element of the development of this system is here to stay and not just in terms of the general sense of human "politics," but actual and sometimes pointed electoral politics of the kind that seems likely to be familiar to most everyone. A political advertisement for a candidate for the

View south along Wilmore Reservoir from its northern end. *Author's collection.*

office of county commissioner in Somerset County raged against the notion that efforts at job creation were being traded away for tourism, totaling up the costs of "tourism" as $78 million, whereas the county had only watched business and enterprise leave the county or go out of business entirely. It was summarized under the pithy phrase of "incumbent commissioner insanity."

Other candidates for office would laud the transformation of the purpose of the system and include it in election bids and platforms. Benson Borough Mayor Tracy Fuller threw her full support behind the possibilities—economic and otherwise—presented by people traveling to take advantage of the boating, canoeing and rafting opportunities along the Stonycreek River, which runs through Hollsopple and Benson.

The focus on the system and of the system was now firmly on its place in local culture and as an end in itself, rather than an instrument to help make steel. This was true even more so than at the mid-century mark, when some started seeking retrospectives and artistic renderings of the reservoirs, especially the Que. There was an increasing awareness of its new purpose and importance: 100 years after the Que's construction, for example, a local historian would lecture and show important photos of its construction to attendees. This same local historian, George Spangler, had a strong personal connection to the Quemahoning—his grandfather was the dam's caretaker for a quarter century, and his own father had that role for another 30 years. Even still, Harvey Stahl was another local watercolorist who painted the Que dam in 1937 and who was profiled near his 95th birthday in 1987.

The ongoing political debates about water supplies for municipal drinking water, a shrinking customer base for the other industrial uses and a significant amount of undeveloped surrounding land all cost money. When the dams were an input for a larger production process, this is factored into the costs of manufacturing, a capital asset easily accounted and expensed, even if numbers such as these were not available for public review but were instead a segment of data in a larger but private and purpose-built system. When the dams, reservoirs, woods and streams are the *product*, however, it turns the cost-benefit and income-outlay expenditures to a new direction. Grants, government funds, loans and other finance mechanisms come to the forefront. All of this was well-known from the very beginning, along with public notices of the needs for funding of various projects and the possibilities presented by different uses, including strip mines. With the early orders for water and interest in the resource, there seemed to be little doubt that the water supply was going to be ready and abundant, but the newly formed CSA also noted that there was plenty of capacity to supply more, with only about a quarter of the reasonable usage capacity being put to work. There had to be other, quite literal, outlets.

RELEASING THE RIVERS

A controlled system of release of the water uses water's own mass to be converted into energy to, for example, turn a turbine, which will then power a motor that converts that physical energy into electricity. A still-controlled, if more dramatic, example of the physics of water occurs with a specialized type of boating known as "white-water," generally in a raft but also in kayaks, canoes and other vessels,[92] as a river's current grows stronger and faster with a greater volume of water released over a shorter period of time. In 2005, both of these uses of water, its potential energy and physics were already well under consideration by CSA and other groups, such as one called the Stonycreek Quemahoning Initiative, or SQI.[93]

> *The concept starts with the Stonycreek River and the fact that it's becoming cleaner. One section already is a popular fly-fishery, another offers one of the best white-water rafting and kayaking runs in the East—when the water levels are up. Then there's the Quemahoning Dam, originally a water supply for Bethlehem Steel Corporation now owned by the public and operated by the* [CSA]. *This one-time industrial water supply now offers fishing, limited boating, swimming, picnicking, camping, and hiking. The Que also offers a lot of water, and the CSA is preparing to send more of it down the* [Stonycreek River] *on a regular basis to help the fishery and, perhaps, create commercial white-water rafting opportunities on weekends year-round.*[94]

The article also notes the other spots along the river and in the area that attract different ranges of interest, which the SQI was seeking to unite. The interests included history and heritage enthusiasts, other environmentally interested organizations and the possibility of economic and community development in the area, responding to the possible influx of visitors taking advantage of the new outdoor recreation possibilities. The article closes the reportage by mentioning the approval of the SQI by the Pennsylvania Environmental Council, calling the SQI "the largest collaboration in Western Pennsylvania and maybe [all of] Pennsylvania."[95]

Just as the momentum of the water itself travels with gravity, the events of the new millennium enwrapped with the different groups of users, different stakeholders and at least a partially different set of values than the previous holders. No longer for steel, no longer private enterprise, the river "sojourns" were and are an amalgam of different users with different values, but the clear goal is no longer an instrumental one; that is, the water for Cambria was an instrument to be used to seek its goal of record. Instead, now, the water is itself the goal. Some quoted experts in the canoeing area stated that the Stonycreek River, as a result of the whitewater releases from the Que, is the "equal of any river [in the eastern United States]." Dave Hurst subsequently noted that the periodic releases and the restoration of the flow of the Stonycreek through its valley brings colder, more oxygen-rich water to the fishery and, indeed, the entire biota of the area. All the awareness and the actual increase in the resource usage for this also came with a necessary increase in awareness of safety of the boaters.

As reported by the *Latrobe Bulletin* in May 2005, "From the river's point of view sojourners enjoy spectacular scenery, experience the corridor's rich history and learn about the amazing ecological comeback that the Stonycreek-

Conemaugh-Kiskiminetas system has made over the past 20 years." For the three-day trip mentioned, sojourners would camp along the Conemaugh and then, on the second night, shift their site to the Quemahoning, taking in industrial sites such as what was left of the Pennsylvania Mainline Canal, which had largely ceased operations by 1840. The yearly sojourns had become a mark of consistent pride by the early 2000s, with the events drawing more interest and participants. Increasing the breadth of interest for those attending, lectures along the way included discussions about the local wildlife in the area and also the natural history of the region. Other groups provided volunteer instruction and donated equipment for the trip as well, enrolling another set of volunteer stakeholders taking part in the transformation of the system.[96] Changes were not dramatic but subtle and generally understated, spread around the system.

Wilmore, like its sister reservoirs, carries a high degree of beauty; whether one wants to add the adjective *natural* to it or not is a matter of choice at this point. It does run a spectrum of adjectives and possibilities for a site or venue; one oft-cited use was that as a venue for wedding photos. The consistent upgrades of more recent history also mean that Wilmore was also the first of the CSA bodies of water to receive construction of a handicapped-accessible pier, followed by the Quemahoning. The funding network for these efforts mirrors that of any other part of the program with its diverse number of contributors: a local bank (Somerset Trust), local architectural and construction firm (Kimball & Associates) and government agencies such as the Department of Community and Economic Development.

But, as considered throughout this work, particularly during times of great transition for the system, there are a multitude of stakeholders viewing shared resources through a different lens, and these mark the occasions when the infrastructure itself becomes more prominent in the discussion. This marks a pronounced social shift in the identity and fabric of the area, one from heavy industry to post-industrial mixed-use ecology. In February 2007, the Quemahoning Reservoir began to participate in the Adopt-A-Lake program, making it eligible for placement of specialized fish habitats. Balancing this out was, by the end of that same year, the finalization on the long-heralded plans for a new water pipeline from the Que. The first pumps would be installed just a few months later. Even mixed-use ecology, as we've seen, has to pay the bills.[97] The property has water and outdoor recreation opportunities, certainly, but it also has timber and coal.

An important element of the CSA mission began, and continues, with balancing the interests of different stakeholders. We have witnessed some

Kayaks available for rent at the southern end of Wilmore Reservoir. *Author's collection.*

of the politically based disputes that arose after the counties' successful bid to purchase the MWC, but the political disputes aren't just about public economy. The disputes can be seen as a type of balancing of interests and rights in the various possibilities offered by the ownership of the property in the region. An example of this is that of continued interest by extractive industries—namely, coal mining. By a split vote in 2011, the CSA Board of Directors voted to grant a lease of coal rights to a Somerset-based energy company to conduct surface mining within the area of the watershed. The chairman of the board, who opposed granting the rights, expressed reservations about surface mining in general terms, although he also noted that the lease agreement contained requirements that the environmental damage be minimized. It might seem contradictory to the mission of the CSA to allow surface mining on a portion of the watershed that it holds in public trust, but this is also the nature of the mixed resource that the system is and has always been, as a product of the region and of the economy at large. Again, a techno-natural area still has to pay the bills for its own maintenance. All these areas are revenue streams for CSA, and the possibility of its continued operation is tied to both the economic needs of the region and the very bald-faced fact of safety of the community as the legacy of a system that needs constant monitoring and maintenance for it to not be a significant threat to thousands.

Changes in the engineering for the Quemahoning also take on a new purpose, with these new sets of goals and intentions. Refurbishment or restoration of a water release valve in the Que dam was a new project that would allow the Que to more regularly take part in the whitewater rafting programs for the region, with controlled releases over a series of weekends in the summer that would bring greater numbers of participants to the area. Also described in these same stories are the themes of historical, heritage and cultural awareness and the partnership that CSA developed for the usage of the former Bethco Pines facility: a summer camp called "Summer's Best Two Weeks." The reportage noted that the usage of the facility is now something available to the public, as opposed to being reserved for "Bethlehem managers and their families." A year later, Dave Hurst again underscored these same possibilities and signaled the continued transformation of the system into an ecological and environmental awareness park, keying in on participation by broad swaths of public users. He sums up the happy

An upstream view of the breast of Quemahoning Dam and the spillway, facing the north. *Author's collection.*

complexity of the partnership by stating that "all of this is happening because more than a dozen groups and hundreds of volunteers worked for a decade to clean up the Stonycreek [River] and remove acidic coal mine drainage that has been killing its aquatic life. With restored life comes recreational and economic opportunities."[98] The Summer's Best facilities, steadily improved and innovated with each year, are also used at other times during the year, such as for educational festivals, as it did in 2008, hosting a Westsylvania festival with an exhibit of a lunar rock, planetarium, presentations by the Pennsylvania Game Commission and historical reenactments of the Civil War and French and Indian War.[99] Other events for the site include Earth Day commemoration and celebration, and it was also a site for a display of a moon rock on loan from NASA. The cultural and historical elements of the area take center stage; amid a dwindling population and largely disappeared major industrial base, what's left can somehow still be seen as more: the weight of shared heritage, history and techno-natural beauty.

One is reminded of Psalm 8:2, "from out of the mouths of babes," when it comes to truth-telling about many things, and there is an interesting example of this with regard to recreation at the Que. One Emily Mishler, a student in the North Star School District—which includes neighboring Boswell and Quemahoning Township—penned a letter for the editor expressing her enthusiasm for the upcoming summer:

> *I spend a great deal of time* [at the reservoir] *during the summer....My sister and I love to go kayaking....The water is far from lovely most days, but the ever-present muck is part of what makes the dam so appealing. It is untainted by the plastic-beautifying efforts of humans, so naturally beautiful animals and native plants can populate the surrounding wilderness. The scenery is quite beautiful in its own way.*[100]

The water clarity is, of course, important on any number of levels, and perhaps this blunt assessment might be thought of as a critique despite Ms. Mishler's further context. What is the most interesting about these thoughts, however, is not the water clarity but the perception of the lake and, indeed, the surroundings, as being something *natural* and evoking "wilderness" and decidedly *not* a product of the human hand. Of course, we see that the system was in fact created by human hands and continues to be maintained by the same, but the larger point that Mishler makes here is illustrative of the final transformation that has been effected by the conversion of the system to what it is now. Even with the longtime hunting, fishing and hiking

A view of the Quemahoning Reservoir, facing northward. In the distance, one can see the breast of the dam peeking above the water. *Author's collection.*

that occurred over the century-plus that the system has been in existence, it is now *seen* as a natural formation or, at the very least, the human artifice entwined with nature in a techno-natural system that is different than what it was because the intention surrounding its use has changed in the culture itself. It is likely that Emily Mishler was born long after Bethlehem Steel wound up its operations in the area and perhaps even after the company itself was bankrupt. She sees the system far differently now than how her forbears might have.

And it isn't always about age, either. Dave Hurst—who undoubtedly does have personal memories of the Bethlehem facility in Johnstown and its full operation—expressed a similar opinion when he advised that the name of the Quemahoning be changed. It should, he said, no longer be thought of as a reservoir. From the *Somerset Daily American*: "Reservoirs store water. While still storing water for sale, this impoundment serves an increasing number of other purposes and deserves the name, 'Quemahoning Lake.'" Such a renaming hearkens back to the news of 100 years earlier, calling the water Lake Quemahoning. All of this, coming a century after the Que's completion, comes as part and parcel of another aspect of the CSA's public mission, which is preservation of its role in history, its place in geography and adding to its cultural weight by placing collections of old construction photographs in the hands of local historical societies. Capping things off from a historical artifact standpoint, the CSA also has a display of a historic water meter. The new, however, can also stand with the old. One example is a laudatory estimate from the *Pittsburgh Post-Gazette* from 2011:

> *At the end of a long damp tunnel 90 feet under the Quemahoning Dam in northern Somerset County sits a large plumbing fixture that has raised the level of enjoyment for water sports enthusiasts and promises to increase it even more. It's a custom-made Henry Pratt fixed-cone 48-inch steel valve that weighs five tons. Call it the "Super Spigot." It's the centerpiece of a $1.3 million construction project designed to attract tourists to the Stonycreek Valley and additional businesses to serve them. When the valve is opened, as it was for the first time on May 13, it creates additional whitewater excitement on the Stony, a river that begins in Berlin in southern Somerset County and heads due north to its merger with the Little Conemaugh River at Johnstown. Until the installation of the valve, canoeists, kayakers and rafters had to depend on snowmelt or rain—and plenty of it—to generate enough water to create exciting whitewater rapids. That dependency will continue because the temporary permit issued by the state Department of Environmental Protection allows a maximum of only eight water releases this year. The second release will be done after stabilization adjustments are made to some of the mechanisms of the century-old dam that directs water to the valve. Those adjustments will enable the valve to increase its water release capability from 2621 gallons per second—the amount released on May 13—to 3740 gallons per second when it's completely opened. The dam has plenty of water—12 billion gallons.*

The depiction of this part of the large techno-natural system carries with it some spirited reminiscences from the past in terms of the glowing and detailed description of the machine itself—the size of the valve, the specifications regarding the water, the age of the dam, the vastness of the reservoir and the appreciation of all of the engineering qualities. It also reminds us of the many stakeholders involved, the regulatory state apparatus and an entirely different conception of the users of the resource. The nexus of the history, engineering, current sporting and recreational use, the limited nature of the resource—along with the *planned human deployment of it*—come together in the form of a single, massive, expensive piece of hardware.

By 2014, new phases of control and monitoring over the system came to be considered. Rather than a more manual system depending on physical presence over a system that is, really by any ordinary considerations, a vast undertaking, the CSA began to examine upgrades to the monitoring and control systems. Rather than manual readouts, constant digital readouts could be collected from remote sensors and combined in an IT-based system. In short, it would be possible to dial up a whitewater release with a smartphone app.

The integration of the system into ecological and scenic surroundings and its transformation into a vehicle for environmental stewardship are not limited to water releases for the rafters and kayakers. Hurst and others spotlighted CSA's efforts in reestablishing the American chestnut tree in the area after most of the species had been wiped out by a blight more than 100 years before. Remarking on what the first European settlers would have seen in the area, he wrote: "We can only imagine such forests, vast and trackless, where one of every four trees was a hardwood we don't see today: The majestic American chestnut. References to chestnut-covered ridges abound in the earliest writings about our region along with descriptions of thick carpets of dropped nuts underneath massive, 100-foot-talll trees with trunks 10 or more feet in circumference." Such a sight has also been echoed by noted Pennsylvania outdoor writer Ben Moyer, with his own memories of the wistful expressions by his grandparents at the beauty and plentitude of the American chestnut, which was wiped out by a fungus brought in on invasive species of Asian chestnut. Both Hurst and Moyer note the complex relationship that humans have with the natural world and the damage that can be caused by what was thought to be an innocuous introduction of a new species,[101] a new species of crossbred American and Asian chestnut with resistance to the fungus that wiped out the great stands of the American chestnut. The CSA partnered with the American Chestnut Society to establish this new orchard in the hopes that something akin to the original species might be somewhat restored to its native range. "For within these buds may reside the genetic material that can finally bring an end to the 100-year winter for the American chestnut."[102]

The pathway to renewed relevance, moving from the industrial water supply to site of environmental awareness for the system of course also means continued engagement in the politics of the administrative and deliberative state. As one would expect, politics is not a conversation; it is argument and it is disputation. Sometimes the disputation is about a resolution of a conflict, and sometimes it is a pathway of compromise that advances progress, however slow it may be. At other times, however, what gets exposed is a different set of competing values, and these are far more difficult for participants to find compromise on, as compromise requires common ground. The public trust mission of the CSA now relies on some general and loose agreement on quite literal common ground and common waters, but the specter of the divisive nature of the national political stage can have a significant impact on the operations of the system and certainly on the perception of its role and purpose. By early 2009, the Que water program had generated enough momentum and capacity to meet the future needs of municipalities in the area served by the water line. One county commissioner was quoted as saying that the added capacity is such that municipalities in the county could "pursue" distribution without having to "search" for water. One further notable aspect of this was the recognition, again, of the large number of partners invested in the success of the program that made the eventual $23.4 million project attainable. A later news item on the same topic stated that the 22-mile length of the pipeline was the largest such project ever undertaken in Somerset County. In June, Somerset County held public events with bottles of water labeled with their provenance of being from the Quemahoning and the new pipeline.

As an example, the Que and its sister reservoirs, along with the entire system, now represent a departure from an industrial past. In a sense, they become barometers for the measurement of political pressure and a gauge to show some degree of progress away from that common heritage. Water quality studies in the watershed point toward the differences between the past and present. In December 2013, a resident of Boswell published a letter to the editor of the *Somerset Daily American* objecting to the spending of $500,000 for new gates for the Quemahoning Dam and some additional spending to repair the current gates. The writer noted: "In these tough times I can see no reason to use our tax money to facilitate a water release so someone can ride their butt down the Stonycreek River in a canoe. Money could be better used to help the many needy people who recently lost their jobs in the area with no hope of finding comparable employment."[103] This is not an unusual refrain in the Johnstown area or, indeed, on the national stage—virtually anywhere that has lost a major economic basis, such as industry or heavy manufacturing, lost thousands of jobs and all of the knock-on economic effects in the area, shrinking and aging populations, a dwindling tax base, greater suspicion and perhaps even anger when actions not only fail to align with values but, indeed, appear to flout them. This perspective is not the only voice in the public-letter sphere; other letter writers would, from time to time, congratulate local officials on their efforts in the area, noting a degree of courage and vision exhibited by the county commissioners in the face of political risk.

The possibility of a hydroelectric power system of some type as a part of the CSA network continued. Several sites were later identified around 2008 and further considered in 2011 that could house such a facility. After considering the Border Dam and South Fork sites in 2008, the power generation would be located at the Quemahoning, Hinckston and Peggy's Run, a small stream but with a powerful downhill movement as a tributary to the Little Conemaugh, precisely through the hillside behind the site of the massive Franklin facility that had been the pride of Cambria Steel. The plant at the Que alone was projected to provide 1 million watts. This has not yet materialized.

The Once and Future Que

The year 2011 was seen as a bit of a—pardon the pun—watershed moment for the system. By this time, the CSA had hit its stride in terms of administering and maintaining the system, the network of partners throughout the region was fully enrolled in CSA and other activities and the publics surrounding them all had a greater degree of awareness as to the opportunities and advantages offered. An editorial by the *Somerset Daily American* hailed it as a "true success story," weaving together the CSA; the elements of the advocates for angling, boating and recreation; and the increase in efforts to ameliorate acid mine drainage throughout the area. Dave Hurst—by this time a well-known and longstanding staunch advocate for the efforts—agreed, reciting all of the technical projects, such as the whitewater release efforts, boating opportunities and fishing habitat to emphasize it. In January 2012, he also took some time to enumerate the different funding streams and interest groups, such as the Benscreek Canoe Club, for their role in these improvements.

Len Lichvar published a piece that gives an excellent recap of the progress made in the ecological and conservation direction for the region, with the usage of the system as a measurement of improvement and an argument advancing values that are distinct from purely economic concerns. It is worth considering in much of its entirety, as it ties together the history of the previous usage of the watershed, a description of the different parties—

public and private alike—involved in the maintenance and turnover of the system and the implicit message of carrying different values forward. It also largely sums up the state of the watersheds now, as we continue the slow march of decreasing pollution and environmental conservation as an end in itself.

Lichvar began his analysis with a smaller project, a part of the ongoing efforts at the Que, such that it readily hearkens back to that 1944 article out of Pittsburgh, heralding Bethlehem Steel's early efforts at acid mine remediation. Through this connection, this continuity, Lichvar states that part of the construction of the dam intentionally altered downstream flows, and the process of the CSA takeover from Bethlehem and the MWC was part of the state government's process for issuing permits to the new owners; part of the conditions of the permit was that stream flows would need to be restored to what they had historically been, prior to the construction of the dam. A large part of the reasoning behind this was to increase the viability of cold-water fish habitat, as previous studies by both the state fish commission and researchers from the California University of Pennsylvania had determined that no trout existed in the downstream watershed. Lichvar noted that, in addition to the state agencies and universities, a number of private organizations also contributed—organizations advocating for cleaner watersheds in the state; organizations for better fishing habitat; local hunting, fishing and conservation clubs; and those with similar affinities.

> *If successful this water will be the state's newest tail water trout fishery and help complete an entire watershed restoration effort that began over 20 years ago. The completion of the Jenners and Boswell AMD* [acid mine drainage] *abatement Passive Treatment Systems restored the water quality of the upper watershed and the Quemahoning Reservoir.*[104] *The public acquisition of the reservoir opened up a huge amount of formerly private land and water to the public and with the upcoming restoration of the lower watershed additional recreational and economic opportunity will be added to the mix. These past and present projects can all trace their roots to the efforts of public sector agencies such as the District, PA Department of Environmental Protection, federal Office of Surface Mining and the Natural Resources Conservation Service, partnering with groups such as* [the Stonycreek-Conemaugh River Improvement Project, the Mountain Laurel Chapter of Trout Unlimited], *Southern Alleghenies RC&D* [resource conservation and development council] *and Conservancy, Somerset County Conservancy, and local watershed and sportsmen's groups.*[105]

That same day, another local group was honored for its work at improving a neighboring watershed along Dark Shade Creek, another tributary to the Stonycreek River that had suffered significant environmental damage from acid mine drainage. The Shade Creek Watershed Association garnered recognition from the same chapter of Trout Unlimited; the association monitors treatment systems and also began to successfully pilot a program for the stocking of brook trout in a nearby stream.

The impetus of the acid mine drainage continues. Starting 2014 off with a vigorous footing, a *Somerset Daily American* headline blared: "Abandoned Mines Pollute Local River." The contrast it offers also stems largely from the economic and regulatory considerations that are spelled out for a system of active wastewater treatment to directly rehabilitate the watersheds, in contrast with some of the passive systems mentioned above. The coal company of concern in the piece, Rosebud Mining, had taken over a site from the Berwind-White company and extended its mining permit to include usage of the previous systems along with a commitment to establish a multimillion-dollar trust fund to continue to operate the

A sign at the breast of the Quemahoning Dam, announcing the fisheries goals for the area watershed. Also notable here is the explanatory text regarding species of interest and the brands and logos of the various partners that are involved. *Author's collection.*

waste facilities even after Rosebud ceased operations at that site. The entire purse that Rosebud expected to lay out was in excess of $30 million, with the invariable contrast between the resources to be spent by a willing private enterprise and those of a sometimes partnership with public groups that do not have the great wells of resources or funding. Just as in 1916 and 1944, most of the environmental focus in the piece was centered on the attainment and maintenance of a fishery on the Little Conemaugh; a source quoted for the article discussed the heavy amount of acid pollution draining into the Conemaugh River, contrasting it to the situation with the Quemahoning, the water releases from which created, in his view, a legitimate and sustainable fishery along with the rafting and boating opportunities. The goal of having a sustainable fishery in the area seems to be the refrain that resonates the most with the policymakers, officials and representatives of the concerned public with regard to pollution costs and the exercise of clean-ups.

It is perhaps somewhat obvious why that affinity would occur; the angling public has the most interest in clean streams and available habitats to fish. In addition, conservation groups tend to have at least some influence on elected officials and administrators alike, as they represent significant user populations that are more easily tracked and measured, through license and stamp sales, through survey responses reported back to the Fish Commission and so on. It is quantifiable and easier to recognize than other measures, such as letters to the editor or even arriving in the constituent inbox at the local elected officials' offices. Perhaps more notable than that obvious hypothesis is the demographic shift in the area in its understood and accepted usage—not as a water supply per se, especially not an industrial one, but instead the view of the system as a "natural" resource unto itself. The same shift also reifies the networks of partnerships among individuals and groups, as the Fish Commission itself hosted mentored fishing events at Wilmore, Hinckston and Quemahoning reservoirs.[106]

The balance continues and the pendulum swings. A later article by Mark Nale in the *Centre Daily Times* related his first-person experience rafting on the Stonycreek. He mentioned 1990, in particular, as the year in which *all* the rivers were dead. Through the use of the initiatives (mentioned in greater detail earlier) and the grants—along with the ever-present economic reality of the price tag—the whitewater releases and the continuing usage of the Stonycreek, the Que and its related system all point to a new vision of resource "exploitation."[107] No longer by primary role an extractive site, it is now a participative site with new groups of participants now publicly using the resources in a somewhat different way—with, all the while, the understanding that these resources had been used for outdoor activities for the preceding century as well.

A Steady Work in Progress

In the summer of 2013, Tecumseh Redevelopment entered an application for a permit to begin work on remediation of the site of the coking facility near Hinckston. The legal scope of the work was to be under the Pennsylvania statute called the Land Recycling and Environmental Remediation Standards Act, one of several statutes among different states and the federal government for the recovery and remediation of brownfields. Tecumseh also had a more recent engagement at another Bethlehem-oriented site in Erie County, New York, at the former Lackawanna facility. A later disclosure showed a bond release for a company performing reclamation work on the Tecumseh property in 2020, showing that the original permit had been granted in late 2006. We see then the legacy of de-industrialization and newer generations of policy intertwined throughout not just the area, but across state borders as well.

Hinckston also has been the site of ecological events. The reservoir, about one-tenth the size of the Que, still has a watershed area of 10.8 square miles, and the water itself, deeply ensconced in a high-walled valley, is 416 acres with 1.1 billion gallons stored. Of the three main reservoirs in the system, it is by far the closest to Johnstown, which, despite its shrinking population, is the most populous nexus in the region. As we've seen in this work, the Hinckston Run area has also long been the site of fishing derbies and hunting and shooting events, dating back to more 100 years, evidently since Cambria Steel relaxed its exclusionary policies.

Closer to more current times, the ecological events include outdoor religious services, weather-related guest lectures and talks, photography expeditions, children's activities and historical reenactments and commemorations. Events like Accuweather Days were sponsored outings for children who ordinarily would not have been able to experience outdoor activities such as fishing to Hinckston. At the same time, groups including the CSA continued to build trails around the reservoir, following paths originally cut by logging roads. Community-service groups and individuals helped cut the paths, showing enrollment of still other groups, including the leadership in this case of the Laurel Highlands Historical Village, working in concert with CSA and others. The CSA does not own all of land surrounding the Hinckston waters. It engaged another group of partners and participants, along with some of the consistent ones, such as the Pennsylvania Game Commission, which donated dozens of bluebird houses for placement around the reservoir. Affiliated groups continued to do the same around the Que as well, as seen in May 2016.

The battle over and with infrastructure takes many shapes and forms. In this work, it's largely been conceived as, above everything else, a possible looming threat. Lack of maintenance here isn't a bridge collapse or a sinkhole—it would be to kill dozens, hundreds, perhaps even thousands if the "right" catastrophe occurs. But the larger issue of infrastructure also takes shape in the form of the ideas captured on paper or electrons or orally communicated. Meetings generate minutes, and the deeds of property exist for hundreds of years, an infrastructure conferring ownership and providing proof of the same. Scientific test results are stacked in the records, with numbers showing levels of pollution or improvement away from that state. Maintenance records of rainfall and pressure, herbicide application and contracts and grants and reports all show social and economic agreement, which provides the invisible skeleton of ideas making up the substructure for the superstructure. Many of the records and types of information involving upkeep and safety are collected and tabulated in charts noting things as various as masonry cracks, vegetation growth and pipe condition. Correspondence involving the service of financial instruments such as bonds locks in the obligations for the supply of resources that must be concentrated in contracting with partners to perform the physical work involved. It is only in this context and this system today that such things can be considered or even conceived, and it is also a reminder of the fundamental oddity that is a techno-natural system. One might be forgiven for looking at a lake in the woods and thinking it was always there, but the reality is that these lakes, in this system, are the outcome of a process involving ink, paper, engineering, litigation, argument, mathematics, commerce and thought.

The enrollment of other entities in the network of interests borne by the system is also significant and established. There are a number of continuing projects with the system, many of which continue to be focused on public access to activities reliant on clean water.[108] The institution of a program for whitewater releases gained broad public support and was seen as both an economic possibility for outside user groups of the system but also local engagement acting as the key to its success. In 2017, the *Somerset Daily American* reported on two parallel events occurring in the coming days: with the support of the Pennsylvania Fish and Boat Commission, the anticipated placement of 40 new fish habitat structures in the Quemahoning Reservoir and also the anticipated "Special Fishing opportunities" program involving Wilmore, Hinckston Run and the Quemahoning, where youth anglers are allowed to keep panfish (bluegills, crappies, perch) that have not yet achieved the minimum size as normally required under the regulations.[109] Programs like this are aimed toward enhancing awareness and usage of the outdoor activities as well as increasing the awareness of the role of regulating hunting and angling in wildlife and natural resource management. Programs such as these also depend on public and private sources of funding and, of course, volunteer labor. Such programs would also continue for years and are continuing to this day, indicating a degree of success of the program, along with other special programs such as reduced-cost Labor Day fishing licenses. At the same time, truly large game fish, such as northern pike, can be caught from the system as well.[110] There have been many newspaper reports over the years commemorating the notable catches of anglers, in addition to straightforward stocking plans and news relating to the management of these fisheries, despite the putative exclusion of the public at large.

The lakes have participated for years in the state's Big Bass Waters program. A later report favored Wilmore's crappie bite,[111] and this panfish action would continue for years. Other reports include stocking of trout at Hinckston Run and bonanza-type catches out of the Que after a heavy rainfall. Eventually, the CSA would continue to use its motor restrictions on watercraft (electric motors only) but would relax its length restrictions, no longer requiring vessels to be less than 17 feet. The striking aspect, however, is the continued focus and emphasis on the water as an angling resource.

It has also enrolled other users during its time. While the transition of the system from Bethlehem to CSA has largely been about public uses, recreational applications and natural conservation, along with natural resources usage such as limited coal mining or lumber, the essential purpose for which Cambria Steel originally built the system is still in place—the use of water for industrial purposes.

In 2016, CSA executed an agreement with Competitive Power Ventures to supply water to a natural gas-powered electricity generation operation near Hinckston Run. CPV has since turned into the largest

Hinckston Run reservoir. *Author's collection.*

single customer of CSA. Other industrial water users include legacy steel enterprises still in the area, including Gautier Steel and Liberty Wire, both current owners of the former physical plants owned by Cambria and then Bethlehem Steel. Other customers include drinking water suppliers, such as the Greater Johnstown Water Authority and Somerset. A power company near Ebensburg also draws on the water supply. North American Höganäs also uses water from the CSA in order to make their alloy products.

The continuity of the use of the resources stands in contrast to the perception of the use of the resources. Throughout everyday conventional wisdom, Bethlehem Steel did not approve of the public use of the reservoirs for conservation purposes. And yet, we see consistent such usage over decades or even a century or more. In considering this in retrospect, it might have been difficult indeed to completely exclude an entire population from the usage of what looks so much like a public good, despite the nature of it being private property. At the same time, there is a life-saving resource at stake in some cases as well, with the use of the system as a supply of drinking water. This use is something that clearly has come about more in recent decades, perhaps, than at the outset, though this is likely also a phenomenon that can be seen as the political wrangling over the usage of the reservoirs as a resource held in trust for public use.

The history of these waters is deeply enmeshed and entangled with some of the most profound, sordid and triumphant periods in history. The construction of the dams represents that triumph, and their usage today stands triumphant as well, with the CSA monitoring this exemplary and complex resource for many diverse and sometimes diametrically opposed stakeholders. Clearly, the events surrounding the Rosedale evictions, the 1937 Little Steel Strike and even the naming of Hinckston Run itself stand as legacies etched in blood from which the new system must continue to emerge in order to atone and build a new history based on common usage and enjoyment.

Another aspect of the techno-natural system is the "connection" to other projects with similar, or at least aligned, goals of environmental awareness and conservation. Communities in the area seek funding and resources to create connections to the system, as Jenner Township did in 2006, seeking resources for the extension of a trail to connect to the Quemahoning area. A strictly technical system used for industrial purposes provides a necessary central pivot, a center of gravity for other engineering solutions to meet challenges that the administrators of a system encountered. The reframed and recalibrated techno-natural system creates a similar platform. It takes a diverse combination of stakeholders with enough of a common vision to create something of this nature.

The CSA has long had budget items for emergency repair work. As noted throughout, this LTS, like any other, requires constant maintenance and upkeep, so the financial resources for not just new items like boat launches and fish habitats. Pipelines, even as well constructed as they were at the outset, crack and break and need repairs.[112] Additional income from the ecological *and economic* possibilities represented by the LTS system is an important consideration, as it was in 2012, when added revenue from logging operations was immediately deployed for anticipated emergencies to repair the pipelines or other related work. By 2012, the role of the CSA system could be firmly understood in terms of its hunting, fishing, and conservation mission. All of this would occur amid continuing efforts to provide water to customers and sell new customers on the possibility of that water service.

Many of the maintenance items are not wholesale, capital expenditure types of repairs and changes, but the smaller amounts do accumulate. A report from 2017 can give a bit of a flavor as to what the CSA faces:

- *$8,767 for fixes near Hinckston and Border Dams;*
- *$1,622 for Que and Border Dam work;*
- *$1,200 for work on a valve near the Hinckston Run.*

This page: Restoration of the historic Keim cemetery near the Quemahoning Reservoir and the Stonycreek River. This was conducted as an Eagle Scout project and continues to be maintained. *Author's collection.*

Breast and spillway of the Quemahoning Dam. *Author's collection.*

The largest expenditure from that list was nearly $24,000 for the Quemahoning Pipeline. The maintenance work isn't just about leaking pipelines and valves, however, so much as it is an ongoing concern for the coexistence of a techno-natural system situated amid history and current politics.

For example, the CSA pays rent for its offices and has forest stands that need to be managed, and both represent outflows of cash. In addition, the system and its management require attention to its debt service and other financial matters, as those are intangible but no less important infrastructural supports. Participation in community affairs and events also means that the CSA is expected to be a contributor, whether in-kind exchanges and partnerships with heritage groups such as the Laurel Highlands Historical Village or a contribution to a fundraiser. It all matters, and it all adds up. In addition, CSA also takes opportunities, when finances allow, to purchase additional property if it adjoins or enhances the current system, as it did with a small tract of land containing a tributary to the Wilmore Reservoir in 2018. With profound, focused and dedicated help from human hands, the waters abide.

6

SOME CONCLUSIONS

Maintenance writes the enduring story of civilization. Without the unsung acts of upkeep—tightening bolts, patching code, revising laws—our best machines and institutions are doomed to rust and irrelevance. Maintenance isn't a chore; it's the quiet engine of continuity. When maintenance takes its rightful place beside innovation, we're not just preserving what we have: we're safeguarding the very foundation of civilization.

The above is from an organization called The Long Now Foundation, a group of thought leaders considering existential risks and the necessity of long-term thinking. There is perhaps no better example of the importance of maintenance than a dam. The dams and reservoirs built by Cambria Steel have outlived their maker. They've also outlived the successors of the maker, including, at one time, one of the largest steel companies in the world. They've outlived hundreds of thousands of people in the area, and we should all recognize that, barring catastrophe, they will—and should—outlive us all. They are a tribute to what might be termed, in one sense, the audacious engineering of nature and matter and for some an example of something that "just can't be done anymore."

The crucial exception to this is that they *must* be done every day—planting grass, tightening bolts, adding new technology to ensure their robustness and safety. They are then too, with this, both the examples of and results of power, the epicenter of struggle for some and the casting forward of ourselves into an ever-changing present and future. The LTS that they represent is itself evolving, having outlasted so many institutions and people, but it must evolve in order to maintain its silent duty: to remain, to stop, to save.

What they save now is different from what they did when first built, at least insofar as their putative purpose and design. What were billions of gallons of cold water to quench the flames of molten steel, which, at least in part, protected the nation, are now billions of gallons of water to preserve habitat, conserve wildlife and allow humans a few precious places of quiet and serenity. And yet, as we have seen, these same artificial bodies of water have always been seen as a source of naturalness itself, for as soon as they were built, it seems, humans flocked to them—to hike, to fish, to see something other than mines and smokestacks.

Vinsel and Russell stated, "You don't run an infrastructure agency and remain free from criticism." This is certainly true in the case of the CSA, as it is for any institution that would be in the position of allocating the resource and balancing them in the interest of all stakeholders that might be present. One of the largest stakeholders, the creator, is, of course, gone now. Others, all of whom had no say in the creation of the dams, are present in the shadow of the dams and are both the witnesses and beneficiaries of their presence now. Arguments about use will always remain; one argument, however, that can never be lost, is that of maintenance. The old axiom is that those who do not learn from history are doomed to repeat it; this is, perhaps, an oversimplification for any number of reasons too complicated to dive into within these pages, but one also need not know the exact history to see its importance.

And what of the dams today?

Reviewing the annual consulting engineers' reports gives a clear picture into the types of maintenance work regularly undertaken for the system. I have included the most specific information from the most recent (2023) report available, though a truly thorough and in-depth review would go year by year, with an ongoing cataloging of the changes made. That may be of interest for another scholar to undertake, but for the purposes here of understanding the many transitions of the system over time, that level of detailed review is not necessary. Instead, the 2023 report reviews the ongoing areas of each dam and reservoir, along with a summary of past improvements and the stated intentions for forecasted and anticipated repairs, remediation and other actions to be taken for the preservation and upkeep of the dams.

For Wilmore, the assessment is that the dam is in very good condition, with routine checks of the integrity of the spillway and the dam breast. Several high-water events occurred over the years, some of which are mentioned elsewhere in the discussion in this work; it's also relevant to know that Wilmore, when first purchased by Bethlehem Steel in 1960, underwent a substantial renovation and capital improvement financed by the steel company. Because of its location directly upstream from a town and its location on the Little Conemaugh—the river that flooded Johnstown in 1889—there is a higher degree of scrutiny and likely fear about it. Nonetheless, the latest assessment of the dam indicated that the maintenance plan and execution by CSA are congruent with the needs of the dam itself. The operators of the system include routine inspections of the dam, including visual inspection by divers and video recordings, which were also reviewed.

For Hinckston Run, the consultants assessed it as being in "very good" condition. At the same time, one of the several unique features of the dam/reservoir is that Bethlehem Steel, and likely Cambria before it, used the dam as the location to dump slag—waste from the steel making process—against it, piling thousands upon thousands of tons of this mass against the dam over the span of decades. Slag is, basically, rock consisting of silica and other minerals leftover from the smelting (steel making) process—it's what's left of the ore. After decades of moving the slag to the Hinckston site, the dam has been essentially buttressed by the many thousands of tons of "debris." The assessment by the engineers is that a catastrophic failure of Hinckston is "extremely remote" due, in part, to the slag reinforcement. If we were all to disappear and maintenance were suddenly stopped on the system, Hinckston is the least likely to suffer a break; it will, as is the case with all lakes left to time, eventually fill in on its own.

For the Quemahoning, this large and historical crown jewel of the system, the dam is as robust as it has ever been. It's also the gateway to new dimensions of conservation and usage, the likes of which were likely never contemplated during the time of Bethlehem ownership. While the company may have tacitly countenanced trespass usage of the resources for fishing or hiking, whitewater releases were never on the agenda. It is now, thanks to the CSA and large swaths of the public, united through sometimes disparate stakeholder networks. The work to rehabilitate the purity of the waters continues as well, for the Que and for the others. Instead of a new strain of history, however, we have seen that this concern has persisted for a century or more and Bethlehem Steel itself was the first to experiment with the possibilities.

An organization such as the CSA, with the large techno-natural system that it administers, is constantly interwoven through a continuous dynamic from past, through the present, to beyond the foreseeable future. Two recent news items encapsulate this well; one piece summarizes the CSA's release

Area of new growth of riparian forest at the Hinckston Run Reservoir. *Author's collection.*

Riparian forest and aquatic life at the Quemahoning Reservoir. *Author's collection.*

of a series of new videos to social media and content site YouTube, with messages of awareness regarding outdoor activities that can be performed at different spots around the system. Another article states that the CSA "continue unearthing water leaks along a...section of the Quemahoning Pipeline." Battling wintertime weather, the painstaking work unearths the massive pipeline, spotting more than a couple dozen small leaks that needed to be repaired. The constant need for such repairs should not be a surprise to anyone, nor are they new—and nor are the potential for disputes, as one can rightly anticipate.[113]

The work is, however, constant. Thus the ongoing need to bring awareness and usage to the system, saddled with the legacy of maintenance and repair of a system more than 100 years old, with not only just catastrophe prevention, but to continue to provide the goods and services to a diverse user community and to somehow connect them all. The outdoor activities are, in a sense, another continuity since the system began. We have seen now for 100 years or more that while the water itself was used for a specific industrial purpose, the whole of the system has been used by many others for reasons found *with* the water. Fishing and boating (when legally allowed, of course) come readily to mind, as do the contemplative aspects of being near to both heritage and natural beauty.

The need for maintenance of the system is necessary as a central component of it, perhaps even making for an identity of its own, especially given its context in the Johnstown area, with its legacy of a flooded and tragic past. The maintenance is also a form of continuity, a handoff of duty from generation to generation, with previous generations tending it for the making of steel and quenching of fire and current generations tending it for the purposes of sublime natural beauty, wise and balanced resource use, public enjoyment and environmental stewardship. The character of the system is radically different in one sense—its utilitarian purpose—along with its set of user groups. Other user groups, as we've seen, represent that continuity since the system's inception. The maintenance also carries through in the physical world to prevent—or clean up after—vandalism. As long as there is some contingent of humans who are willing to improve a situation, there are groups of people willing to attempt to undo something helpful and constructive. What possesses the minds of some to do this will forever elude most of us who appreciate the hard work involved in keeping something nice.

Maintenance isn't necessarily in only the physical realm, either. As noted elsewhere, the CSA has needed to use legal processes to enforce contract provisions, but in addition to the law, other knowledge economy– and service sector–based maintenance can occur as well. In other instances, this occurs in the realm of finance, not only in the sense of income and expenditure but in other measures such as bonds to secure funding for the physical work or measures of refinancing bonds as well, which better the economic conditions and parameters under which the system may run. On occasion, maintenance and upkeep come in under budget, which means less expenditure and therefore a lower financial burden on the taxpayer, as CSA's annual loan amounts from Somerset and Cambria Counties, on occasion, can decrease. The costs can be high, however, but the availability of water itself, along with the recreational opportunities the system creates, can justify the expenditure from the standpoint of the policymaker.

As we have seen, perhaps surprisingly, there seemed to be a long-term awareness of environmental damage by industrialization. Prior to 1920, discourse included concern about runoff pollution by tanneries. In the midst of World War II, a major daily in the steel-making center of the country discussed concerns about acid mine drainage. By 2013, with now more than a decade of experience with the CSA, there had been

A well-known local landmark of a waterfall flowing from Hinckston Run Reservoir. *Author's collection.*

A view from the eastern shore of Hinckston Run Reservoir, looking west-northwest. The pads visible on the surface are turtle sunning platforms, installed by the Pennsylvania Game Commission, one of the crucial stakeholders in the ongoing effort to preserve this important natural resource. *Author's collection.*

measurable improvement in the area watersheds, including the Que. At the same time, the increase in efforts to create more and better fish habitats also continued; in the view of Len Lichvar, quoted in this piece and elsewhere, it was more about the restoration of that which had been lost as a result of industrialization rather than anything "new." One could then ask—is this a new continuity or perhaps the recovery of a continuity that existed prior to an interruption *by* Cambria Steel Company when they first built the system?

For the system to continue in this continuity-focused way for the next 100 years, stakeholders and participants will need to continue to unify themselves around that common-enough vision for conservation, environmental and recreational uses. That may, at times, abut the need for the descendants of the heavy industry still in the area to pave economic pathways via the extraction of coal or the managed sale of timber stands. There is always an uneasy alliance for things such as this, but the high economic, social and moral value of the system itself will, one hopes, carry the day for that common vision and focus.

NOTES

Introduction

1. The two smaller dams are the South Fork Dam, not to be mistaken for the South Fork dam that ruptured in 1889, and the Border Dam.
2. While there are a number of excellent resources, the two best have stood the test of time. One is the paradigmatic study done by David McCullough, *The Johnstown Flood*, published in 1968, and the other is the Oscar-winning short documentary of the same name.
3. The actual number of persons buried in the plot is, perhaps surprisingly, a bit confused. The first official count was around 751, with additional markers bringing it to 777 for symmetrical purposes. However, much later, a researcher actually counted the cross-shaped markers and came to a total of 813, thereby posing the mystery—was the number of actual dead and unidentified at an unusual 777? Of course, 777 is just as likely to occur as 776 or 778, but the mystique of the number 7 in Christian beliefs as being affiliated with God and celebrated as a lucky or blessed number remains. The matter is still up for debate.

Chapter 1

4. The authors also note that the 1889 flood is still the costliest dam breach in the history of the United States and the deadliest dam disaster.

5. The geological legacy left to Pennsylvania includes hundreds of different types of soils. Soil itself is a combination of ground rock, minerals, water and organic matter and is relatively recent in a geologic sense. There are no existing soils that predate the extinction of the dinosaurs, and most important in considering ecological challenges, a significant majority of the soils today date from the Pleistocene or the most recent Ice Age. Soils, then, are a fragile component of the agricultural web and the ecosystem and, rather than overlooked as "dirt," can more truly be seen as the anchor of and foundation for so much of what we rely on today as far as food, water and the natural world.
6. Reports on the continued viability of the dams are cause for concern; two of the three have been rated as in need of significant repair, discussed later in this work.
7. *Pottsville Evening Republican*, September 7, 1935. Even occasioning a report of a black widow spider, the web of which had captured "a cricket, wasp, bee, and bettle [*sic*]."
8. The Franklin plant is the most obvious example of this phenomenon. The facility had once been the largest in absolute terms in the Cambria Steel arena (irrespective of whether one considers Cambria Steel as a separate entity or as a part of a still larger entity, such as Midvale and then eventually Bethlehem). Franklin was constructed just before 1900 and was itself an integrated steel works, with multiple blast furnaces, rolling mills and coke ovens and also included much of what one would expect—railway sidings, larry cars and many thousands of laborers. The vast stretch of land along the little Conemaugh River in Franklin Borough—the namesake of the plant itself—is now a scarred and nearly barren crater that gives little indication as to the totality of its operations and the thousands of laborers and families that work there once supported.
9. I use the word *larger* here guardedly, mostly in the sense that the system was built with the goal of steel production in mind, but a system consisting of five dams and reservoirs, one of which holds greater than 11 billion gallons of water, along with miles of pipeline and the assorted supporting machinery and other implements, is hard to describe as "small" or even "smaller." This is further emphasized when considering the significant role and risk of the topography itself and its ability to channel large volumes of water into paths of destruction.
10. There is another theory generally named the Social Construction of Technology, or SCOT, which goes into far more detail regarding users of and adoptions of technology.

11. *Pittsburgh Post-Gazette*, June 9, 1919. At other times, the dams themselves become a part of the infrastructure of another large technical system familiar to everyone—roadways and incessant road repair and construction, which, scarcely to be believed, predate the interstate system that Americans know all too well today. Reports from the summer of 1919 abound regarding the usage of the road bordered by the Hinckston Run dam to be a detour into Johnstown during another road closure nearby.
12. Vinsel and Russell, *Innovation Delusion*, 14.
13. Vinsel and Russell, *Innovation Delusion*, 38–39.
14. The Quemahoning is also the general area that, in July 2002, captivated the nation during the harrowing rescue of nine coal miners trapped underground. The mine is located near the Quemahoning Creek, with the Quecreek being the portmanteau of the words; hence the name "Quecreek Mine Rescue" is in the history books for a quiet spot in Somerset County less than a year after and a 10-minute drive from where Flight 93 crashed.
15. The words from Emma Lazarus's poem *The New Colossus*: "Give me your tired, your poor, your huddled masses yearning to breathe free" have become a shibboleth for the narrative of limitless economic opportunity for everyone, irrespective of economic or social status.
16. It is also worthy of note that Morrell was invited to be a member of the South Fork Hunting and Fishing Club, the entity responsible for the negligence of the dam that burst and killed thousands. Morrell joined more in the interest of keeping his eye on things, as his own livelihood and business was downriver. Morrell never saw the flood of 1889 and the damage it wrought; he had died four years earlier.
17. Some of these included the noted "Panics" of 1893, 1896, 1907 and 1910–11. The recessions also tended to be somewhat lengthy, a few of which numbered for several consecutive years. One gets the impression that the economy spent at least as much, if not more, time in contraction or fear of retraction than anything that we would witness in more recent times.
18. While the population loss is frequently cited in any number of opinion pieces relating to the collapse of the manufacturing economy, one can also note that the city and its surroundings are fairly balkanized, with boroughs and townships not factoring into the city population. In short, people in 1920 lived closer to the mills, often within walking distance but also via network of trams and streetcars. As times changed and the

automobile became more dominant, greater numbers of workers lived outside the city proper. Even with that caveat, however, the counties of Cambria and Somerset also peaked in population in the first half of the 20th century and have steadily lost population overall. Somerset County (the location of the Quemahoning Dam) peaked in population in 1940 at just nearly 85,000 and Cambria County (the site of both Wilmore and Hinckston Dams) peaked around the same time with the number at 213,000. The numbers are now, respectively, 74,000 for Somerset and 133,000 for Cambria.

19. Coal companies may not always be the most creative resources for naming the towns formed by and around them. "Windber" is an anagram of Berwind, of the Berwind-White Coal company. North of Windber, in Cambria County, are the twin towns of Colver and Revloc, both settled close to coal mining operations and both charmingly anagrams of each other. At least these towns received names, however; many others in the area only got a number reflecting the number of the mine that they were closest to. Examples of this include Mine 42 and Mine 37, small so-called coal patches between Windber and Johnstown. Even more are the abandoned towns along the Ghost Town Trail in Cambria and Indiana Counties, where dozens of towns once stood and are now gone.

Chapter 2

20. From the collection *Johnstown: The Story of a Unique Valley.*
21. *Somerset Daily American*, April 21, 1983. This has been an oft-repeated origin story throughout the history of the system and particularly the Quemahoning Reservoir.
22. This was also projected to be able to provide 75 million gallons of water per day during a time of drought. The addition of "flashboards" on the spillway would allow the reservoir to retain water over capacity for slightly additional water usage and needs as well.
23. Though not, as we see later in this chapter, without writing very significant checks.
24 An early report from the *Pittsburgh Weekly Gazette*, July 25, 1905, noted that the MWC claimed that the lake would have a capacity of 4 billion gallons; this was a considerable underestimate.
25. *Wilkes-Barre Semi-Weekly Record*, July 3, 1900.

26. *Somerset Daily American*, August 1, 1979, and August 8, 1980. A number of celebrations commemorating Hollsopple occurred over the years, including successive ones in 1979 in an exploration of the history of the borough and its actual 100-year anniversary a year later.
27. *Pittsburgh Daily Post*, March 17, 1906. This is likely not surprising, given the deeply intertwined nature of the topography, the people and heavy industry. By 1906, for example, there was already a tunnel near the Quemahoning and bearing its name, for railroad traffic between Somerset and Pittsburgh.
28. A brief piece in the *Pittsburgh Daily Post*, July 2, 1900, is one of many regional dailies that emphasized the discontent of the farmers and the likelihood of litigation, if not outright violence.
29. Some other sources claim it was $250,000—either way, a substantial amount of money.
30. Rininger made national news; a story with substantially the same language appeared a week or so later in the newspaper in Walla-Walla, Washington.
31. The most expensive average price per acre of land today in the country is in Rhode Island, with an average of $350,000 per acre. Of course, for more urban areas, real estate prices can be astronomical, but these comparisons lend insight into the rather dear price that land was fetching when Cambria Steel was buying.
32. This sum is the equivalent of more than $135 million in 2023.
33. The Church of the Brethren is a Protestant Anabaptist mainstream church with around 600,000 members nationally; there is a significant presence of the church in Bedford and Somerset Counties, Pennsylvania.
34. This same report, from the *Boston Evening Transcript*, noted that the 11-billion-gallon water capacity of the reservoir was enough for a city of 500,000. Perhaps the stringers for the short piece thought that an interesting factoid to give the readership a sense of size, or perhaps they were laboring under the misapprehension that the Que was being constructed for drinking water. If the latter, that was mistaken, at least for the time.
35. *Somerset Daily American*, October 16, 1985. Dibertsville was the site of an early Dunkard church, subsequently covered by the Quemahoning's waters. A later letter contribution to the *Daily American* (March 8, 2006) written by Richard Blough attested to his personal witnessing of foundations of homes owned by his family that had been revealed during time of drought.
36. Cambria Steel Company Field Engineer's Testing Report, December 23, 1913. The test was undertaken on November 18, 1913.

37. *Pittsburgh Daily Post*, April 3, 1917. This policy was rescinded—at least temporarily—as the country edged closer to participation in the First World War, and the Cambria mills were already making vast quantities of war materiel. Cambria Steel began guarding the reservoirs with police powers and constant searchlights to guard against sabotage and a "precaution against possible plots to cripple" the facilities.
38. For those with experience angling or those who have experiences with anglers, it might be understood that the prose in report generates a familiar feeling of a "fish tale" not only from the size of the fish but also of the extensive reporting of the sub-par tackle. August 4, 1914, was also the day that Germany invaded Belgium, irrevocably triggering the entry of the United Kingdom and France into war.
39. One might get the impression that once married, a woman lost her entire name, unless Mary Clawson's sister's birth name happened to be "Mrs. John Warren."
40. The Quemahoning Reservoir would remain the largest lake in the state until the mid-1920s, when it was superseded by the far larger Lake Wallanpaupack, in the Poconos region in the northeast of the Commonwealth.
41. More precisely, the creek supplying the reservoir (the "run") was named as the trout recipient.
42. *Pittsburgh Post-Gazette*, May 6, 1930. Except perhaps in one notable instance: in an ironic change from the threats to dynamite the Quemahoning dam in 1900, an individual in 1930 was detonating dynamite in the Wilmore reservoir to kill a large number of fish and presumably harvest them. He was duly arrested and fined.
43. *Pittsburgh Post-Gazette*, June 20, 2008; *Pittsburgh Post-Gazette*, May 29, 2009. Occasional fishing reports would surface from time to time over the history of the system and the dams. In 2008, for example, Wilmore and Hinckston Run were both noted for panfish, while trolling Quemahoning resulted in "limited numbers of walleye and brown trout" in addition to largemouth and smallmouth bass. A 2009 report notes the abundance of crappies at Wilmore. I fished for crappies at Wilmore in my youth as well, accompanied by the constant and not entirely needed paranoia because we "weren't allowed to fish there."
44. In a strange coincidence of the two events, a report from the *Meyersdale Republic*, May 23, 1918, noted the status of the local boy-done-good, Corporal Fred Emerick, who was in the hospital in Texas prior to deployment to the European theater, in order to have some corrective

surgery on his arm, which had been damaged by a concrete mixer while building the Quemahoning Dam.

45. In 1981, the *Somerset Daily American* ran a 50-year remembrance of the recovery from a 1931 drought that had "lowered the water to a dangerous level."

46. *Chambersburg Public Opinion*, July 23, 1927. Especially prior to World War II, the Quemahoning was a common site for outings and picnics, as in the case of the Over and Gilbert families, who held their reunion there in July 1927.

47. *Allentown Morning Call*, June 28, 1924. The *Boswell News*, 40 years later, would post a "remember 40 years ago" article commemorating the discovery of the body on July 2, 1964.

48. *Meyersdale Republic*, July 27, 1944. In an interesting display of the continuity of normal life during times of war, a laudatory report in July 1944 also announced the presence and message of a noted minister from Chicago coming to visit and to teach at Camp Harmony.

49. *Pittsburgh Post-Gazette*, May 26, 1938. The camp included, by 1938, sporting fields, tennis courts, swimming opportunities and rooming and dining facilities enough for "several hundred people."

50. Coke ovens are the devices by which bituminous coal is combusted to a point where it removes all of its volatile compounds and is left with a matrix of nearly pure carbon fuel, called coke. Coke has a higher level of efficiency for blast furnaces,

51. Cody McDevitt's work provides a significant addition to the ofttimes untold or under-told disturbing history of the banishment of Blacks and Mexicans from Johnstown.

52. *Somerset Daily American*, January 1, 1975; *Johnstown Tribune Democrat*, August 15, 2022. Hinckston was also the site of one of the more infamous crimes in Johnstown history. Given its location—in terms of its proximity to Johnstown but its comparative isolation due to the removal of the Rosedale neighborhood and its operations—it can be the site for both solitude and an escape from watchful eyes. The main road connecting it through the city—now largely an ATV trail—became overgrown and unused. Honan Avenue had neither maintenance nor traffic for years, and the difficulty of access and lack of need to use it made it become a spot where dirty deeds might either occur or be the site where the remnants of the deed were left. In the case of Barbara Mangus, hunters discovered her body around New Years Day 1975, after she had been reported missing two weeks prior. She had been strangled after leaving a Christmas party,

and witnesses reported seeing her in the West End section of town prior to her disappearance. Her homicide remains unsolved and is a distressing and recent remnant of a brutal crime occurring in the twilight of the Steel Age of Johnstown.

53. It did lead to another statewide "Johnstown flood tax" that was put in place and remains so—it is a surtax on liquor purchases in the Commonwealth. In addition, the Roosevelt administration granted a significant public works improvement project to channelize the rivers around Johnstown and alleviate catastrophic flood risk. This was thought to be a vital success, as Johnstown was left largely unscathed by the other statewide and regional flooding that occurred in the wake of Hurricane Agnes in 1972. Five years later, however, the flash floods of 1977 did break several smaller dams around the city, flooding it and killing dozens.

54. *The Indiana Gazette*, March 15, 1980. As is the case with such events, the fact that the death toll wasn't as great as what could have been contemplated led to a feeling of relief, with "only" a few people killed, rather than thousands. However, that is not to say that it was less tragic for the families affected by the death and destruction that was wrought. In 1980, Nick Dutko published a personal reminiscence of his childhood memories of surviving the 1936 flood, noting the fear he felt while struggling through the waters of the flooded Conemaugh River, as his family home was near Cambria City, downstream from both the Little Conemaugh and the Stonycreek. "Unforgettable is the panic that followed," as he put it when he encountered hundreds of people fleeing from the water, shouting that the Quemahoning Dam had broken and those billions of gallons of water were surging towards the city.

55. A brief piece in *Centre Daily Times* of March 19, 1956, noted the 20th anniversary of the 1936 flood, including the rumor that the Quemahoning Dam had burst, flooding Johnstown, and that the rumor had caused great panic.

56. My searches have not revealed what, if anything, Bethlehem Steel or the management of the local facilities (what had been Cambria Steel Company) said before, during or after the flood of 1936. It is unlikely that they made any type of public statement, as they might have been relieved that, in fact, nothing catastrophic occurred, and felt no need to add to the weight of words that had already been placed on top of the water. What I would note is that, as a part of my own family's history, my maternal grandfather first came to Johnstown as a part of

the clean-up efforts and was then hired on at Bethlehem Steel in the aftermath. He finished his career in 1977, retiring as general foreman of the riggers.

57. A *Monongahela Daily Republican* headline on January 22, 1937, blared: "PITTSBURGH AWAITS 33-FT CREST: FLOOD WATERS PUSH WAY INTO RICH TRIANGLE." Interestingly, next to a column on a spate of recent auto thefts, the celebrity buzz of the day: the possible romantic link between Katharine Hepburn and Howard Hughes.

58. On June 20, 1956, the Pittsburgh Post-Gazette reported a brief death notice of one ET Gray, who passed at the age of 74. Gray was the person who journeyed to the Quemahoning Dam on the day of the 1936 flood to confirm whether it was holding or was in trouble. The paper noted that Gray was a "prominent Johnstown civil engineer."

Chapter 3

59. See my work *Johnstown Industry* (Arcadia Press, 2021), for a more detailed overview of the facilities, especially the Franklin facility, which is now largely gone for scrap and is an empty and barren strip of land east of Johnstown. Franklin alone had more than 200 coke ovens, totaling more than 400 for the entire Cambria facilities, which included its own local coal mines (including the infamous Rolling Mill Coal Mine, which in 1902 was the site of one of the deadliest coal mining disasters in the nation's history, where an explosion killed more than 100 miners), 8 blast furnaces, 4 Kelly (Bessemer) Converters, 4 blooming and slabbing mills, 10 rail mills, 13 bar mills, 1 wire and rod mill, 6 laboratories, and 4 of its own proprietary power plants. With this background in mind, it is perhaps little wonder that the complex had its own water supply as well. Even into the 1970s, when the facilities and the entire industry were decades past their peak, there were around 12,000 people employed (out of a city which then had a population of around 40,000).

60. Unsurprisingly, the blasts also made major news in Baltimore, as reported by that city's *Evening Sun* on June 29, 1937: "2 Blasts Halt Steel Output, 6000 Made Idle." Baltimore was, after all, the site of Bethlehem's largest mill, at Sparrow's Point.

61. White, *Last Great Strike*, 52.

62. Discussed in greater detail herein and in Cody McDevitt's work *Banished from Johnstown.*

63. The parallels between his maneuvers and that of Mayor Cauffiel are striking; Cauffiel had done the same 14 years earlier to persecute minorities.
64. White, *Last Great Strike*, 183.
65. After such a cascade of terror embedded in the collective memories of thousands, it is little wonder why areas such as Johnstown and its rural surrounds have a deeply ingrained cultural support for an expansive interpretation of the Second Amendment.
66. The history of the Pennsylvania State Police as a force has a close entwinement with the burgeoning heavy industrialization of the 19th and early 20th centuries. In the late 1800s, policing was done in part by the Pennsylvania Coal and Iron Police, a semi-privatized group of brutish thugs whose mission was to disrupt organized labor action. This force became more and more unpopular—beating people to death in plain sight tends to have that effect—and was disbanded, with statewide law enforcement moving more and more to the Pennsylvania State Police, a professional organization administered by the state government that originated as the State Constabulary. The CIP had been in decline for decades following a 1902 anthracite strike, and the PSP were established in 1905.
67. *Pittsburgh Daily Post*, September 5, 1908. The reader will also recall some of the early threats pertaining to the Quemahoning Dam included it being blown up by angry farmers or frightened residents of Johnstown, dating from 1900. Explosives seem to be a recurring theme, as in a report from 1908 indicating that someone had planted sticks of dynamite in the fuel coal storage bin on a locomotive heading from the Quemahoning area to Cumberland, Maryland, presumably to kill whomever was in the engine of the train.
68. The description in the article noted a "five-minute tug" for the fish and that the young angler would be the "envy" of fishermen everywhere. That a three-pound largemouth bass, while indeed a nice catch in a northern latitude such as that of Hinckston today, made a newspaper 100 miles distant might be considered in the context of what the state fisheries must have been like, overall, at the end of World War II and during the peak of industrial production in Pennsylvania.
69. More on this later, in the later portions of the history of these waters.
70. The article was published in June 1952, so it is somewhat unclear whether the writer, when talking of autumn-painted hills, was using their imagination or memory or if the piece had been held onto by the editors and not published until months after having been written.

71. It's beyond the scope of this work to recount the history of the PRR; an interested reader should go to Albert Churella's comprehensive three-volume history of this massive, storied corporation for the entire story. By the time of the acquisition of Wilmore by Bethlehem, PRR was well into its decline and would, within the decade, be merged with other neighboring railroads. It went bankrupt a short time after that.

72. I note that this is a shift in perception because I am reluctant to determine any type of "shift" in the nature and character of the country itself. This is the role of politics and politics *using* history to reestablish its own goals through a lens of power and desire. I am content to leave the political judgment to others, as any type of persuasion has little point, at least in terms of learning anything new.

73. Nucor is the top steel producer in the United States as of 2021, with eight China-based companies and a number of Japanese, Indian and South Korean producers all outproducing even them. Nucor is, however, nearly 40 percent larger than its nearest United States–based competitor, which happens to be the legacy company United States Steel. USX also is, as of the time of this writing, again a subject of political rhetoric from politicians, unions and management, as Nippon Steel of Japan has entered an agreement to purchase USX. In the steel doldrums of the 1970s, however, Nucor was the upstart pioneer of the so-called mini-mills, which were steelworks of much smaller footprints and labor costs than the massive steel plants that one tended so see around Johnstown, Pittsburgh and other historical areas of heavy industry.

74. This was also the era of the "malaise" speech made so famous by President Jimmy Carter in 1979. With spiraling debt and unemployment rates, an inability to be effective on the international stage, with gasoline shortages and inflation reaching into every household's pocketbook, the 1970s were a difficult time of increasing mistrust in both business and government. With regard to President Carter, in an astonishing fit of accurate assessment, a politician judged the country's position so truthfully and correctly that he failed to gain reelection against Ronald Reagan in 1980.

Chapter 4

75. Like many in the area, the author lost relatives in this flood, particularly with the damage to Tanneryville. Rest in peace, Sheldon and family.

76. For a more extensive review of the 1977 flood and its effect on local steel production, Patrick Farabaugh's work is an excellent resource: *Disastrous Floods and the Demise of Steel in Johnstown*, published by Arcadia Press.
77. *Somerset Daily American*, April 15, 1981; *Somerset Daily American*, January 23, 1982; *Somerset Daily American*, February 5, 1982. At around the same time, there was a proposal for a possible synthetic fuel manufacturing facility using the water supply of the Quemahoning. The proposal received federal funding for a study, but indications are that while those monies were allocated, any further work or support from the federal Department of Energy was likely not coming during the first Reagan administration. In January 1982, the siting of such a facility seemed "probable." That would end of being premature, as within a few days, one of the key challenges with the project was the lack of a market for the methanol to be produced, and finding funding for the actual plant itself was an insurmountable challenge at the time.
78. *Somerset Daily American*, March 19, 1984; *Somerset Daily American*, April 2, 1988; *Somerset Daily American*, March 22, 1997. The usage of the system for outdoor recreation and experiences with the natural world is not new and is one stream of continuity throughout its lifetime. The Audubon Society, for example, sponsored and organized spring waterfowl observations on the Quemahoning Reservoir. Four years later, a number of area clubs and interest groups combined to enhance the habitat of area lakes, including the Quemahoning, by constructing and placing duck boxes (generally a simple wooden box that protects a hen and her eggs) on a number of lakes in Somserset County, including the Que. Waterfowl and other birdwatching would continue throughout the 1980s and 1990s, with one source claiming that the area's birdwatching is the best.
79. *Somerset Daily American*, August 2, 1997. The system continues to contend with a number of risks, with the acid mine drainage risk being one among several. Simple and prosaic risks also do damage and cause eyesores, such as littering or even full-scale illegal dumping. A watershed, by its very nature, is also vulnerable to accidental exposure to risks that we all live with but with effects that multiply beyond the immediate and obvious, as in the case of filling stations or heating oil tanks.
80. *Latrobe Bulletin*, October 27, 1997.
81. *Somerset Daily American*, May 13, 1964. Not all of the business of Bethlehem Steel was private, however. Negotiations regarding tax rates made the news, especially the wrangling over the assessment of the value of Bethlehem properties around the Quemahoning Dam, where the county increased

the assessment by essentially quadrupling the valuation. Bethlehem contested this and counteroffered with a little less than a doubling of the assessment. The county accepted Bethlehem's counterproposal.

82. *Somerset Daily American*, July 7, 2001. It did not take long for the CSA to form its own network of critics. Barely a year after its formation and instituting its policies for the maintenance of the system, a writer and angler wrote a letter to the editor protesting the CSA's rule limiting boats to the use of electric motors only. The writer asserted that the formation of the CSA itself was a "mistake" and that while the purchase of the system was on the whole a net benefit, the ownership should have been transferred to the Pennsylvania Fish and Boat Commission, which tended to allow limited-horsepower gasoline outboard engines for powering watercraft.
83. One could justifiably and conceivably argue that the land-use lawsuits that arose in the wake of Cambria Steel's actions in the early 1900s were likely more inefficient from both an economic use and human resources standpoint than what Bethlehem and other parties did in the late 1990s, in its negotiations with bidders, following regulatory processes and meeting the transparency requirements for public finance. Regulatory and administrative tradeoffs have a rich literature that might point toward some interesting conclusions in this area.
84. *Somerset Daily American*, November 15, 1999. The commissioners of both Somerset and Cambria Counties were given an award for the formation of the CSA and the successful purchase bid of the system.
85. *Somerset Daily American*, September 17, 2011; *Somerset Daily American*, April 16, 1975; *Somerset Daily American*, April 25, 1984; *Daily American*, July 26, 2009. Local health systems and other groups routinely have hikes and runs still, such as the Conemaugh Health System 5K. This usage of the resource has taken a number of different causes over the years, such as awareness for those with intellectual disabilities and impaired vision, among others. The Quemahoning Dam in particular seems to be a popular destination both for the walk and for its end. Other groups, such as the local Lions, host tribute events around the Que.
86. That setup of haves and have-nots in the circumstances plays well for its resonance with the management versus labor dynamic and also the poetic symmetry that can be drawn with the inception of the South Fork Hunting and Fishing Club and the ill-fated dam of the 1889 flood. None of these comparisons are particularly apt or even cogent, but it's a tribute to the enduring nature of an assumption archetype that they tend to get repeated.

Chapter 5

87. These articles are still quite obviously available with an elementary web search; researching the Wilmore Dam, for example, is made somewhat difficult by the presence of hundreds of web links to articles detailing a disastrous flood that never occurred in September 2021.
88. *Standard-Sentinel*, May 1, 1931. "A systematic and thorough examination of dams in the state is made at intervals by engineers of the Water and Power Resources Board in the Department of Forests and Waters."
89. This would, of course, not be without controversy. While there are many dire news pieces about coal mining contamination and risks posed by strip mining, a well-run operation can minimize environmental burden and restore the contours of the original hill at the close of its extraction activities. Still—the consideration of it brings a pragmatic optic to the early days of the CSA and the necessity of funding maintenance.
90. *Somerset Daily American*, July 8, 2009; *Somerset Daily American*, February 17, 2010; *Somerset Daily American*, December 27, 2007; *Somerset Daily American*, August 4, 2010. Being a public entity, the CSA uses public notices for a variety of purposes and reasons, including a bidding process for the removal and salvage of the original 66-inch stainless steel pipe. Such a process would occur in public at a set place and time, where bids would be unsealed and announced. In 2010, CSA published another solicitation for bids, which included a pre-bid meeting in the same physical location in order to help assure consistency of responses to the solicitation, which sought a partner for the upgrade of some sections of the water pipeline. Similarly, awarding of contracts also received news attention. From time to time, CSA has its public meetings at the reservoirs themselves, such as at the breast of Wilmore Dam.
91. *Daily American*, May 7, 1966; *Daily American*, May 8, 2016. The reader may recall an earlier note regarding firefighters battling blazes in neighboring Bedford County in 1931. The region isn't particularly known for having large wildfires on a regular basis, as opposed to what one might witness in the western part of the United States. Even still, however, they do occur and are, perhaps, even more frightening due to their rarity, as was the case in May 1966, when widespread fires were reported around Somerset County, including the area around the Que. This was later recounted as an anniversary event fifty years later: "Tinder dry forests became the prey of flames in what was described as one of the worst forest fire situations in the county's history."

92. *Bloomsburg Press-Enterprise*, May 20, 2018. Even stand-up paddleboards, for the most adventurous types.
93. *Somerset Daily American*, December 1, 2006. A significant number of syllables for a three-letter acronym. The initiative continued to evolve in a number of areas, including the Quemahoning fishing tournament, discussed elsewhere. A 2006 piece by Dave Hurst also noted the trans-disciplinary interests of the initiative, bringing both interests in natural conservation and history together.
94. *Latrobe Bulletin*, October 26, 2005.
95. *Somerset Daily American*, June 28, 2004. The ongoing process for finding the funding to run the "new" system started at the very beginning and continued as the system developed. In 2004, for example, then–State Representative Tom Yewcic assisted in procuring grant funding from the Pennsylvania DCNR for promotion of the natural beauty of the area and for assistance in upgrading the Quemahoning Dam's capacity to perform whitewater releases.
96. *Somerset Daily American*, February 17, 2010. During the funeral of longtime Congressman John Murtha, he also received credit for continuing to provide resources to the Somerset County area for the support and maintenance of the Quemahoning pipeline.)
97. *Somerset Daily American*, April 8, 2015; *Johnstown Tribune-Democrat*, May 12, 2021. And sometimes, more than just the regular ones; the CSA ended up suing one of the contractors for pipeline work in 2015 over a dispute concerning the work's specifications in the agreement. The need for maintenance is complete and never-ending, however, as the CSA also solicited bids for on-call emergency repair contractors throughout the entire system.
98. *Latrobe Bulletin*, September 9, 2009; *Somerset Daily American*, January 22, 2004. Dave Hurst continued this steady drumbeat through a series of regional articles that he authored and published, such as one in September 2009, again noting the pending improvements to the Que dam and its contribution to the whitewater releases and the ecological and heritage significance of the area and, implicitly, the presence of this large technical system. Hurst's drumbeat for the greater Stonycreek watershed, the Quemahoning and the dam stretches back for more than two decades prior to the writing of this book, including early pieces heralding new life—both economic and environmental—in the area.
99. *Johnstown Tribune-Democrat*, August 23, 2013. While reenactments were held for demonstrations, there was no known engagement in either war

actually near the site. On other occasions, contributions by locals to the Civil War effort would be recounted in learning experiences offered at Hinckston as well.

100. *Somerset Daily American*, May 9, 2013.

101. Moyer also remarked on the recent—and somewhat underreported—blight on ash trees wrought by the emerald ash-borer, an insect that has virtually destroyed all species of ash in North America during the past couple of decades.

102. *Somerset Daily American*, January 6, 2011. Hurst also took on a leading role in the promotion of boating in the "Stonycreek Canyon," as he called it, heralding the beginning of the resurgence of the river leading up to the work to replace the release valve in the Quemahoning, along with the new area tradition of kayakers and canoeists meeting up on January 1 for a cold-weather paddle down the river.

103. *Somerset Daily American*, September 7, 2011; *Daily American*, September 8, 2011. Larry Williams had written a similar sentiment two years earlier, that "our elected officials could be more focused on solving the problems of the taxpayer rather than the comfort of a few riding their butt in a canoe down the river. I guess our thrill is going down the river with them without a canoe while drowning in debt." This letter was one to draw a strong and acidic response from the president of the local canoe club, correcting the record for the distribution and collection of the funding (with the majority of it being raised via private means) for the Quemahoning efforts.

104. *Pittsburgh Post Gazette*, October 14, 2002. Ironically, the rescue of the Quecreek miners in July 2002 included pumping large amounts of water out of the flooded mine, which released significant acid runoff into the watershed. This is perhaps the most brutal reminder of the tradeoffs involved with ecology and heavy industry; human lives at risk from the way in which the people make their living, combined with the circumstances of an accident and the justifiable and frantic efforts to save the lives involved can cause environmental harm. It would have been unconscionable to let the miners meet their fate without any attempt to rescue them; it would be nonsense to suggest otherwise or that the rescue itself caused the environmental harm—especially when the circumstances were such that the miners were doing a job they had been hired to do and, in some cases, had been doing for decades. It is just the result of a sometimes-uneasy coexistence; as we have seen in the history of this system, many of the employees of the steel mills and coal mines availed themselves of

opportunities to hunt, fish, hike and bike through the same areas that were potentially at risk through the nearby industrial activities.

105. *Somerset Daily American*, March 30, 2000; *Somerset Daily American*, April 20, 2002. The Southern Alleghenies group also hosted fundraisers with the backing of private businesses such as car dealerships, which often featured tests and contests of fishing skills, such as a casting contest. Another example of a volunteer group tending to the improvement of the system: the Mountain Laurel Chapter of Trout Unlimited sponsored a clean-up near the outflow of the Quemahoning Dam.

106. *Somerset Daily American*, October 28, 2014. Interestingly, the remembrances of the usage of the reservoirs and the property continued, despite some of the confusing assessments regarding the exclusive "ownership" by Bethlehem. In the run-up to Halloween in 2014, a brief ghost story written by Donald Hagans—*The Phantom Beagle*—tells of the author's encounter with a ghostly canine around 1960 while hunting near the reservoir.

107. *Somerset Daily American*, May 1, 2003; *Bloomsburg Press-Enterprise*, May 20, 2018. The requests via grants began early in the life of CSA as well, with a feasibility study for the regular water releases awarded a grant in 2003. In addition, there can be an interesting question for consideration here, in that the whitewater conditions are created on a schedule, rather than constant, and that the schedule is designed to attract users and tourists for some distance with the re-created whitewater conditions for recreation are reflective of an authentic "natural" experience; it is, however, a question that is largely without meaning, given that the system is now a "techno-natural" one that is unlikely to be removed given economic, environmental and sociological considerations. Whitewater on a regular schedule is better than a catastrophic flood, after all, and there are very few areas that are truly untouched by human influence. In fact, there may not be any.

108. *Somerset Daily American*, April 8, 2010. And the system can symbolize stewardship and awareness at a fundamental level as well, such as being a site for anti-littering "Adopt a Highway" type of approach.

109. *Pittsburgh Post-Gazette*, July 7, 2002; *Somerset Daily American*, August 8, 2014; *Somerset Daily American*, August 20, 2015; *Somerset Daily American*, August 23, 2017. At first glance, this might appear counterintuitive or counter to the ordinary goals of allowing smaller fish to grow, similar to antler restrictions on deer. It is the case, though, that some species of fish—such as bluegills—are prone to population surges, which can result

in a drastic reduction in available food in the aquatic ecosystem, harming the quality and quantity of the fish available for anglers. This approach has been offered at many times over the years, including at the three lakes of this system. Other notices for special fishing opportunities and news of fish habitat improvements are quite frequent as well.

110. The *Somerset Daily American* of May 27, 2017, shows an angler with a 2-foot-long pike on the anniversary of her brother's catch of an even larger 37-inch pike, 22 years earlier—presumably when the Quemahoning was still under Bethlehem's exclusive ownership. Indeed, Alvin Wiencek was pictured with his catch of a 31-inch northern pike as well in the *Somerset Daily American*, September 1983. Even earlier, 10-year-old Susan Miller was pictured with a 2-foot-plus northern pike (September 13, 1980). A later feature in the *Daily American* noted the success of a local angler catching multiple pike (April 15, 2011). Other notable features included a pair of twin brothers successfully consolidating on their spring turkey hunt in May 2006 (*Somerset Daily American*, May 6, 2006).

111. The author notes his own childhood experiences of fishing for crappie at Wilmore to be exceptionally productive, and these were the days prior to CSA's ownership and management of the system.

112. *Somerset Daily American*, March 7, 2013. The system itself is affected when its partner systems also have maintenance issues; the fact that the human-made items need repair and sometimes replacement can both be understood and be registered as an ongoing area of debate and dispute, as was the case with a water line breakage that flooded nearby properties.

Chapter 6

113. *Somerset Daily American*, April 8, 2015. Incomplete or subpar work in the field of repairs is an area of eternal ripeness for contract law, litigation and disputes. CSA has had to resort to legal measures to enforce contracts for repairs as well.

BIBLIOGRAPHY

Akron Beacon, 1903.

Altoona Tribune, 1900–1916.

Anaconda Standard, 1900.

Auburn (NY) Citizen, 2021.

Baltimore Evening Sun, 1961.

Baltimore Sun, 1903–1908.

Bedford Gazette, 1907.

Berger, Karl, MD. *Johnstown: The Story of a Unique Valley*. Collection published by the Johnstown Flood Museum, 1985.

Brooklyn Daily Eagle, 1900.

Brusatte, Steve. *The Rise and Fall of the Dinosaurs: A New Study of a Lost World.* Mariner Books, 2018.

Buffalo Commercial, 1914.

Buffalo Enquirer, April 17, 1905.

Cambria Somerset Authority, www.cambriasomersetwater.com.

Cambria Steel Company Product Handbook, 1909.

Canonsburg Daily Notes, April 3, 1936.

Carbondale Leader, December 11, 1922.

Centre Daily Times, June 14, 2015.

Chambersburg People's Register, February 1, 1907.

Chambersburg Public Opinion, November 5, 1963.

Connellsville Daily Courier, October 16, 1937.

Correspondence, between Gulf Oil Company and Bethlehem Steel, 1984, regarding an oil spill in the Wilmore Reservoir watershed, on file at Cambria Somerset Authority archives.

Coughenour, C.L., N.M. Coleman and A.L. Taylor. "In the Shadow of the Dam—Hydrology of the Little Conemaugh River and Its South Fork, with Insights about Past and Future Flooding." *Heliyon* 8, no. 9 (2022).

Cumberland Evening Times.

Donehoo, George P. *Indian Place Names and Villages in Pennsylvania: With Numerous Historical Notes and References.* Sunbury Press, 2014.

Duncannon Record, April 5, 1934.

Elizabethville Echo, 1900.

Engineering Record.

Engineer Reports, Cambria Somerset Authority Archives.

Farabaugh, Patrick. *Disastrous Floods and the Demise of Steel in Johnstown.* The History Press, 2021.

Field Engineer Report, November 1913, Cambria Somerset Authority Archives.

Francis, Robert E. "A Biography of COL John Hinkson: Pennsylvania and Kentucky Frontiersman." https://frontierfolk.net/ramsha_research/jhinksonbio.html.

Fulton Democrat, April 2, 1936.

Greenfield Daily Reporter, June 29, 1937.

Harrisburg Evening News, 1921–1937.

Harrisburg Intelligencer Journal, January 4, 1928.

Harrisburg Telegraph, 1900.

Hazleton Plain Speaker, June 29, 1937.

Hazleton Standard-Sentinel, December 4, 1930.

Hazelton Standard-Speaker, 2021.

Indiana Democrat, 1900.

Johnstown Tribune-Democrat, 1990–2023.

Juniata Sentinel, July 18, 1900.

Lancaster Intelligencer, February 6, 1907.

Latrobe Bulletin, April 21, 1937.

Long Now Foundation, mass email, December 23, 2023.

McCullough, David. *The Johnstown Flood.* Simon & Schuster, 1987.

McDevitt, Cody. *Banished from Johnstown: Racist Backlash in Pennsylvania.* The History Press, 2020.

Meyersdale Republic.

Mount Union Times, 1900.

Moyer, Ben. *Smoke to See By.* Catamount Press, 2023.

New York State Environmental Clean-up At Former Bethlehem Steel Site, NYSDEC.

Pittsburgh Daily Post, 1904–1917.

Pittsburgh Post-Gazette, 1900, 1910.

Pittsburgh Press, 1937–1961.

Pittsburgh Weekly Gazette, July 24, 1905.

Pittston Gazette, June 29, 1937.

Ponca City News, March 18, 1936.

Potter Enterprise, July 18, 1900.

Pottsville Evening Herald, June 29, 1937.

Reading Times, April 1, 1903.

Sanitary Survey of the Allegheny River Basin, Department of Health, Commonwealth of Pennsylvania, 1915.

Shamokin News-Dispatch, 1945–1963.

Somerset Daily American, 1900–2022.

Southern Alleghenies Conservancy: Report Gathering Public Opinion, August 1999.

The Spectator: Yearbook of the Johnstown, Pennsylvania High School, Class of 1935.

St. Joseph News-Press, September 2, 2021.

Stockton Daily Evening Record, March 18, 1936.

Tyrone Daily Herald, November 6, 1973–1980.

Vancouver Sun, 1936.

Vinsel, Lee, and Andrew Russell. *The Innovation Delusion: How Our Obsession with the New Has Disrupted the Work That Matters the Most.* Crown Currency, 2020.

Warren, Kenneth. *Bethlehem Steel.* University of Pittsburgh Press, 2009

White, Ahmed. *The Last Great Strike: Little Steel, the CIO, and the Struggle for Labor Rights in New Deal America*. University of California Press, 2016.

Whittle, Randy, *Johnstown, Pennsylvania: A History, Part One: 1895–1936.* The History Press, 2005.

Wilkes-Barre Semi-Weekly Record, 1900.

Wisconsin Rapids Daily Tribune, June 29, 1937

York Gazette and Daily, March 20, 1941.